Studies in Computational Intelligence

Volume 1273

The series "Studies in Computational Intelligence" (SCI) publishes new developments and advances in the various areas of computational intelligence—quickly and with a high quality. The intent is to cover the theory, applications, and design methods of computational intelligence, as embedded in the fields of engineering, computer science, physics and life sciences, as well as the methodologies behind them. The series contains monographs, lecture notes and edited volumes in computational intelligence spanning the areas of neural networks, connectionist systems, genetic algorithms, evolutionary computation, artificial intelligence, cellular automata, self-organizing systems, soft computing, fuzzy systems, and hybrid intelligent systems. Of particular value to both the contributors and the readership are the short publication timeframe and the world-wide distribution, which enable both wide and rapid dissemination of research output.

Indexed by SCOPUS, DBLP, WTI AG (Switzerland), zbMATH, SCImago.

All books published in the series are submitted for consideration in Web of Science.

Lakhveer Kaur · Pushpendra Kumar
Editors

Neural Networks: Theory, Algorithms, Simulation, and Applications

Springer

Editors
Lakhveer Kaur
Department of Mathematics
Jaypee Institute of Information Technology
Noida, Uttar Pradesh, India

Pushpendra Kumar
Dhirubhai Ambani Institute of Information and Communication Technology
Gandhinagar, Gujarat, India

ISSN 1860-949X ISSN 1860-9503 (electronic)
Studies in Computational Intelligence
ISBN 978-3-032-18749-9 ISBN 978-3-032-18750-5 (eBook)
https://doi.org/10.1007/978-3-032-18750-5

This Springer imprint is published by the registered company Springer Nature Switzerland AG
The registered company address is: Gewerbestrasse 11, 6330 Cham, Switzerland

Dedicated to My Grandparents

—Pushpendra Kumar

Contents

Node-wise Variable-Order Fractional Diffusion on Graphs: Toward Fractional Graph Neural Networks

Suvendu Nayak and Arun Kumar Gupta

Abstract This chapter presents a symbolic framework for diffusion on graphs with heterogeneous time-varying memory. We introduce a variable-order (VO) fractional Laplacian $L_{\alpha(t)}$ that assigns a local, time-dependent order $\alpha_i(t) \in (0, 1)$ to each node, coupling graph topology with node-specific memory. On the analytical side, we show that the resulting operator is linear and ℓ_2-bounded, and we discuss conditions under which the associated diffusion model is well-posed. On the modeling side, we illustrate the flexibility of the framework through symbolic examples drawn from performance networks, adaptive sensor systems, and memory-aware simulation settings. Beyond physics-inspired diffusion, the framework links naturally to machine learning: replacing the classical Laplacian in graph-based propagation rules with $L_{\alpha(t)}$ yields memory-aware message transmission and suggests a route toward *Fractional Graph Neural Networks* (F-GNNs). In such architectures, the orders $\alpha_i(t)$ can be treated as fixed, scheduled, or learned from data, providing an explicit mechanism to encode heterogeneous memory in networked systems. We discuss data-driven estimation of $\alpha_i(t)$ and extensions to multilayer and temporal networks within this symbolic framework.

1 Introduction

Fractional calculus has become an important mathematical framework for modeling systems with memory and hereditary effects [1, 2]. In contrast to traditional integer-order models that imply essentially instantaneous responses, fractional models allow the evolution of a system to depend on its entire past, which leads to more realistic descriptions of viscoelastic materials [3], anomalous diffusion [4], biological processes [5] and many more [6–11]. Building on this, variable-order (VO) fractional calculus introduces an additional degree of flexibility by allowing the order of differ-

S. Nayak (✉) · A. K. Gupta
Department of Mathematics and Applied Statistics, School of Applied Sciences, KIIT Deemed to be University, Bhubaneswar, Odisha, India
e-mail: suvendu.nayak1601@gmail.com

L. Kaur and P. Kumar (eds.), *Neural Networks: Theory, Algorithms, Simulation, and Applications*, Studies in Computational Intelligence 1273,
https://doi.org/10.1007/978-3-032-18750-5_1

Table 1 Notation

Symbol	Meaning
$G = (V, E)$	Undirected graph with node set V and edge set E
A	Adjacency matrix of G
D	Degree matrix of G
$L = D - A$	Standard graph Laplacian
$x(t) \in \mathbb{R}^n$	State vector of node quantities at time t
$\alpha_i(t)$	Time-dependent memory order associated with node v_i
${}^C D_t^{\alpha(t)}$	Variable-order Caputo fractional derivative operator

entiation to depend on time, space, or internal states [1, 12]. Such VO formulations are particularly suitable for describing adaptive, evolving, and nonstationary systems in which memory depth itself changes over time (Table 1).

In parallel, graph theory has developed into a central tool for analyzing connected systems in fields ranging from social and transportation networks to biological interaction graphs and communication infrastructures [13, 14]. In graph-based models, the Laplacian operator encodes the underlying topology and governs diffusion, synchronization, and consensus dynamics [13]. However, classical graph diffusion models typically assume homogeneous dynamical behavior across nodes and do not explicitly account for heterogeneity in memory or temporal response patterns.

Recent studies in complex systems indicate that real-world networks are often heterogeneous and time-evolving in their effective dynamical behavior. In adaptive sensor networks, for example, individual sensors may experience different environmental conditions and thus exhibit different effective memory depths [15]. Likewise, heterogeneous graph-structured systems such as biological or omics networks exhibit multi-scale response patterns and structural variability [16]. Fixed-order diffusion models are generally not well suited to capture such simultaneously local and temporally evolving heterogeneities.

To better reflect these effects, fractional diffusion models on graphs have been proposed in which the diffusion operator is generalized to a constant or globally varying fractional order [17]. While such models extend classical diffusion, current formulations are often tied to specific numerical schemes and emphasize discretization aspects. They typically offer limited symbolic interpretability and do not provide a general framework for node-wise, time-varying memory profiles.

These considerations motivate the development of a symbolic variable-order fractional Laplacian operator $L_{\alpha(t)}$ that enables each node to evolve according to its own time-dependent fractional memory order $\alpha_i(t)$. A symbolic formulation preserves structural information about the diffusion process and supports analytical investigation, interpretability, and modular extensions [18]. Symbolic approaches are particularly relevant in applications that demand transparency and explainability, including autonomous systems, cognitive robotics, edge AI, and critical infrastructure networks.

In this chapter, we introduce a graph-based symbolic modeling framework in which variable-order fractional derivatives are defined node-wise, while diffusion is mediated by the graph Laplacian. The resulting symbolic VO-Laplacian operator synthesizes ideas from fractional calculus, graph theory, and symbolic abstraction to represent memory-adaptive diffusion on heterogeneous dynamic networks.

Beyond theoretical diffusion modeling, the framework also connects to contemporary machine learning. Graph neural networks (GNNs) have emerged as powerful architectures for learning on graph-structured data, with Laplacian-based operators playing a key role in message transmission and representation learning. Incorporating fractional operators into GNNs has recently been explored as a means of enriching expressive power and robustness by capturing long-term dependencies and nonlocal dynamics [19, 20]. In this context, replacing the classical Laplacian with the symbolic VO-fractional Laplacian $L_{\alpha(t)}$ suggests a natural route toward *Fractional Graph Neural Networks* (F-GNNs), in which node-wise memory profiles $\alpha_i(t)$ can be fixed, scheduled, or learned from data. This perspective places the present framework at the intersection of mathematical physics, network science, and emerging memory-driven neural computation.

The main contributions of this chapter are as follows:

- A symbolic node-wise variable-order fractional Laplacian on graphs that couples graph topology with heterogeneous time-varying memory profiles is defined.
- Analytical properties of the proposed operator, including linearity, ℓ_2-boundedness, and conditions ensuring well-posedness of the associated diffusion model under standard assumptions are discussed.
- A generalized symbolic VO-fractional diffusion equation that captures heterogeneous memory-driven dynamics on networks is formulated.
- Symbolic application examples involving adaptive sensor networks, performance-style dynamics, and decentralized memory-adaptive systems are presented.
- The connections to data-driven learning of memory profiles, integration into graph neural network architectures, and the modeling of multilayer and multiplex networks are discussed.

The rest of the chapter is organized as follows. Section 2 introduces the mathematical preliminaries, including graph Laplacians, variable-order Caputo derivatives, and node-wise memory dynamics. Section 3 presents the symbolic VO-Laplacian operator. Section 4 examines its key mathematical properties. Section 5 develops the generalized VO-fractional diffusion model and its physical interpretation. Section 6 illustrates the framework through symbolic application examples. Section 7 discusses broader implications and limitations and Sect. 8 concludes the chapter.

2 Preliminaries

The following sections outline the mathematical notations and operators behind the symbolic variable-order diffusion model on graphs. We first present a brief overview of graph Laplacians, introduce variable-order Caputo fractional derivatives, and proceed with the concept of node-wise memory dynamics.

2.1 *Graph Laplacian*

Let $G = (V, E)$ be a simple, undirected graph with n vertices. Its adjacency matrix is

$$A = (a_{ij})_{i,j=1}^{n}, \qquad a_{ij} = \begin{cases} 1, & \text{if } (v_i, v_j) \in E, \\ 0, & \text{otherwise.} \end{cases}$$

The degree of vertex v_i is

$$d_i = \sum_{j=1}^{n} a_{ij},$$

and the degree matrix is

$$D = \operatorname{diag}(d_1, \ldots, d_n).$$

The standard graph Laplacian matrix is given by:

$$L = D - A.$$

The graph Laplacian is a matrix capturing structural information and playing a fundamental role in operators controlling diffusion, random walks, and synchronization effects throughout networks [13, 14]. Its eigenvalues and eigenvectors provide meaningful information regarding graph connectivity, clusterability, and speed of diffusion.

2.2 *Variable-Order Caputo Derivative*

Fractional derivatives generalize classical differentiation by introducing derivatives of non-integer order. The Caputo form of fractional derivative is particularly popular because it accommodates classical initial conditions [1, 21]. Furthermore, the variable-order Caputo derivative permits the order of differentiation to be time or space dependent, incorporating changing memory behaviors.

For a sufficiently smooth function $f(t)$ and a continuously differentiable order function $\alpha(t) \in (0, 1)$, the variable-order Caputo derivative is given by:

$$^{C}D_t^{\alpha(t)} f(t) = \frac{1}{\Gamma(1-\alpha(t))} \int_0^t (t-\tau)^{-\alpha(t)} f'(\tau)\, d\tau,$$

where $\Gamma(\cdot)$ is the Gamma function.

The order $\alpha(t)$ can change based on external factors, internal system states, or other dynamics, offering a powerful modeling tool for adaptive memory processes [12, 22].

2.3 *Symbolic Node-Wise Variable-Order Memory Dynamics*

In networked systems, nodes might have various memory behaviors based on their role, context, or internal dynamics. In order to account for this heterogeneity, we introduce a symbolic node-wise variable-order derivative operator.

Let $x(t) = [x_1(t), x_2(t), \ldots, x_n(t)]^T \in \mathbb{R}^n$ be the node quantity state vector at time t. Every node v_i evolves in accordance with its own local memory order $\alpha_i(t) \in (0, 1)$. The symbolic node-wise variable-order derivative takes the form:

$$D_{\alpha(t)} x(t) = \begin{bmatrix} ^{C}D_t^{\alpha_1(t)} x_1(t) \\ ^{C}D_t^{\alpha_2(t)} x_2(t) \\ \vdots \\ ^{C}D_t^{\alpha_n(t)} x_n(t) \end{bmatrix}.$$

This symbolic expression enables each node to dynamically change its memory depth over time, offering a basis for heterogeneous adaptive diffusion modeling. The symbolic structure also supports analytical study, symbolic-numeric hybrid simulation, and modular extension to multilayer and multiplex network environment [17, 18, 23].

3 Symbolic VO-Fractional Laplacian Operator

We now introduce the core construction of this chapter: a symbolic variable-order (VO) fractional Laplacian operator that captures heterogeneous memory-adaptive diffusion processes on graphs.

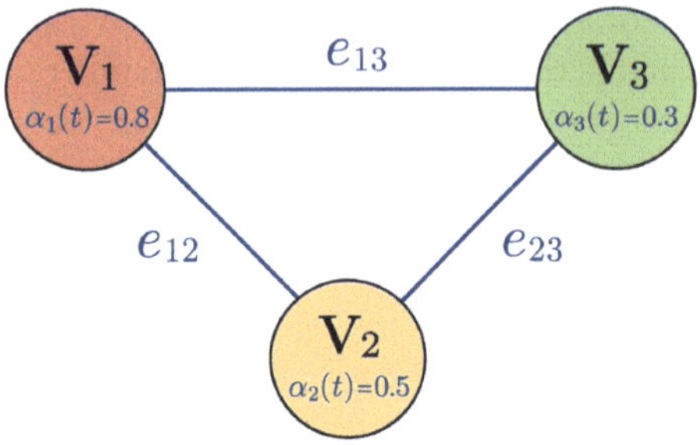

Fig. 1 The symbolic VO-fractional Laplacian operator

3.1 Operator Definition

The **Symbolic VO-Fractional Laplacian Operator** is defined as:

$$L_{\alpha(t)}x(t) = L\,D_{\alpha(t)}x(t),$$

where $D_{\alpha(t)}x(t)$ is the vector consisting of node-wise variable-order Caputo derivatives.

The following Fig. 1 illustrate a schematic example of a simple graph with node-wise time-varying memory orders $\alpha_i(t)$ demonstrating the symbolic VO-fractional Laplacian.

Every node v_i evolves according to its own, time-varying memory order $\alpha_i(t) \in (0, 1)$, whereas the Laplacian structure regulates the spatial diffusion within the network.

The operator $L_{\alpha(t)}$ captures two layers of dynamics with different characteristics:

- **Spatial Dynamics**: Regulated by the graph Laplacian L, determining how information, energy, or influence spreads between nodes.
- **Temporal Memory Dynamics**: Driven by local memory profiles $\alpha_i(t)$ at each node for heterogeneous and time-evolving retention of previous states.

This mathematical model describes actual phenomena in which distinct agents, sensors, or elements have personalized memory behavior in their interactions using a shared network structure [15–17].

The symbolic VO-fractional Laplacian operator generalizes many traditional models:

- If $\alpha_i(t) = 1$ for all i and t, then $L_{\alpha(t)}$ simplifies to the conventional graph Laplacian operating on the first derivative $\dot{x}(t)$.
- If $\alpha_i(t) = \alpha$ constant $\forall i$, then $L_{\alpha(t)}$ is a homogeneous fixed-order fractional diffusion operator on the graph [17].
- If $\alpha_i(t)$ is time-invariant across nodes but varies node-wise, the model embraces spatial heterogeneity in memory but not temporal adaptation.

3.2 *Motivation for Symbolic Representation*

Symbolic modeling, rather than numerical approximation alone, facilitates:

- Analytical study of structural characteristics like linearity, boundedness, and stability.
- Modular incorporation into symbolic AI and cognitive computing architectures [18].
- Interpretable representation of heterogeneous dynamic systems without compromising mathematical transparency.
- Scalability to more sophisticated systems like multilayer, multiplex, and temporal networks [23].

The symbolic representation of $L_{\alpha(t)}$ therefore offers an open and transparent platform for representing adaptive memory-based dynamics on complex networks.

4 Mathematical Properties of the Symbolic VO-Fractional Laplacian

In this section, we derive important mathematical properties of the symbolic variable-order fractional Laplacian operator $L_{\alpha(t)}$, namely linearity, well-posedness of the corresponding diffusion model, and boundedness of the operator. Linearity and boundedness of $L_{\alpha(t)}$ support the analysis of stable energy propagation in memory-based diffusion processes, and well-posedness guarantees that the system evolves deterministically under suitable conditions. These attributes are essential for employing symbolic VO-diffusion models in adaptive systems such as sensor grids, cognitive networks, and decentralized learning environments [17, 18, 24].

Theorem 1 (Linearity) *The symbolic VO-fractional Laplacian operator $L_{\alpha(t)}$ is linear with respect to the state vector $x(t)$.*

Proof Since the graph Laplacian L is linear [13, 14] and the variable-order Caputo derivative operator $D_{\alpha(t)}$ is linear [21], their composition $L_{\alpha(t)} = L \circ D_{\alpha(t)}$ is also linear. That is, for any $a, b \in \mathbb{R}$ and functions $x_1(t), x_2(t)$, we have:

$$L_{\alpha(t)} (ax_1(t) + bx_2(t)) = aL_{\alpha(t)}x_1(t) + bL_{\alpha(t)}x_2(t).$$

□

Theorem 2 (Well-posedness) *Consider the initial value problem for the symbolic VO-fractional diffusion model:*

$$L_{\alpha(t)}x(t) = F(t, x(t)), \quad x(0) = x_0, \tag{1}$$

where $F : [0, T] \times \mathbb{R}^n \to \mathbb{R}^n$ is continuous and Lipschitz in $x(t)$. Then, under the additional assumption that the operator equation is equivalent to a well-defined variable-order fractional evolution problem on the admissible state space, there exists a unique local solution $x(t)$.

Proof Sketch. Under the stated additional assumption, the problem can be treated as a well-defined variable-order fractional evolution problem. The local existence and uniqueness result then follows from standard fixed-point arguments under continuity and Lipschitz conditions on $F(t, x(t))$ [12, 21]. □

Theorem 3 (Boundedness) *For each admissible state $x(t)$, the operator $L_{\alpha(t)}$ satisfies the ℓ_2-norm bound*

$$\|L_{\alpha(t)}x(t)\|_2 \leq \lambda_{\max}(L)\, \|D_{\alpha(t)}x(t)\|_2,$$

where $\lambda_{\max}(L)$ is the largest eigenvalue of the graph Laplacian L.

Proof Because the graph Laplacian L is symmetric positive semi-definite with real non-negative eigenvalues [13], its operator norm in the ℓ_2-norm satisfies

$$\|Ly\|_2 \leq \lambda_{\max}(L)\, \|y\|_2, \quad \forall y \in \mathbb{R}^n$$

where $\lambda_{\max}(L)$ is the largest eigenvalue of L.

Taking $y = D_{\alpha(t)}x(t)$ yields

$$\|L_{\alpha(t)}x(t)\|_2 = \|LD_{\alpha(t)}x(t)\|_2 \leq \lambda_{\max}(L)\, \|D_{\alpha(t)}x(t)\|_2.$$

□

5 VO-Fractional Diffusion Model and Physical Interpretation

The symbolic variable-order fractional diffusion model on a graph is expressed as:

$$L_{\alpha(t)}x(t) = F(t, x(t)),$$

where:

- $L_{\alpha(t)}$ is the symbolic VO-Laplacian operator as defined earlier,
- $x(t) \in \mathbb{R}^n$ is the state vector at time t,
- $F(t, x(t))$ stands for external forcing, control, or environmental impacts on the system.

The initial condition is given by:

$$x(0) = x_0,$$

where x_0 is the initial state.

This model encapsulates diffusion phenomena in which each node stores memory adaptively through its own time-dependent memory order $\alpha_i(t)$.

The VO-fractional diffusion equation can be physically interpreted as describing systems in which:

- **Heterogeneous Memory Effects**: Different nodes remember the past with varying intensities and time-evolving profiles [15, 22].
- **Adaptive Diffusion Dynamics**: Diffusion strength and persistence are locally and temporally adaptive.
- **External Modulation**: Inputs or perturbations affect node dynamics in addition to intrinsic memory evolution.

Systems described by this model include adaptive sensor networks [16], developing biological networks, cognitive networks, and decentralized robot swarms, in which memory retention influences coordination schemes.

When $F(t, x(t)) \equiv 0$, the steady-state equation is given by

$$L_{\alpha(t)} x_{\text{ss}} = 0.$$

If the memory profiles $\alpha_i(t)$ turn time-independent (i.e., $\alpha_i(t) = \alpha_i$ constant), the steady-state behavior can be treated similarly to classical diffusion models, but with memory-corrected dynamics [3, 4].

The variable-order memory radically changes convergence rates, equilibrium distributions, and system stability properties.

A symbolic formulation offers several benefits for modeling and controlling future complex systems that need to run autonomously, adaptively, and transparently. These benefits are:

- Explicit node-wise control and tracking of memory evolution.
- Analytical derivation of bounds, stability conditions, and equilibrium properties.
- Easier integration into symbolic AI systems and cognitive computing frameworks [18, 24].

6 Applications

In this subsection, we exhibit the modeling flexibility of the symbolic VO-fractional Laplacian framework through three exemplary demonstrations: a cricket-inspired performance network, a distributed adaptive sensor network, and symbolic simulation frameworks.

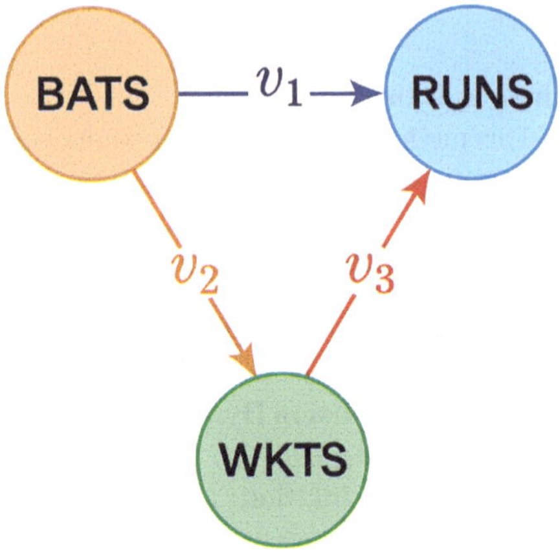

Fig. 2 Symbolic cricket-inspired performance dynamics network

6.1 Cricket-Inspired Performance Dynamics Network

We simulate a reduced-order cricket performance network where nodes represent important match dynamics such as BATS (batting form), RUNS (runs scored), and WKTS (wickets lost). The edges represent influence between these performance attributes (Fig. 2).

In this symbolic VO model:

- BATS possesses long memory, capturing the cumulative batting performance and psychological self-assurance.
- RUNS possesses intermediate memory, changing faster in response to recent patterns of scoring.
- WKTS possesses short memory, responding nearly instantaneously to wicket loss and team upset.

The symbolic VO-Laplacian encodes heterogeneous memory profiles dynamically affecting momentum changes, collapse threats, and strategic interventions during cricket matches [16, 17]. These symbolic insights into performance dynamics set the stage for broader applications in adaptive sensing, where heterogeneous memory patterns also play a critical role.

6.2 Distributed Adaptive Sensor Network

We represent a smart city sensor network, where sensors vary their memory behavior according to environmental variability and physical location (Fig. 3).

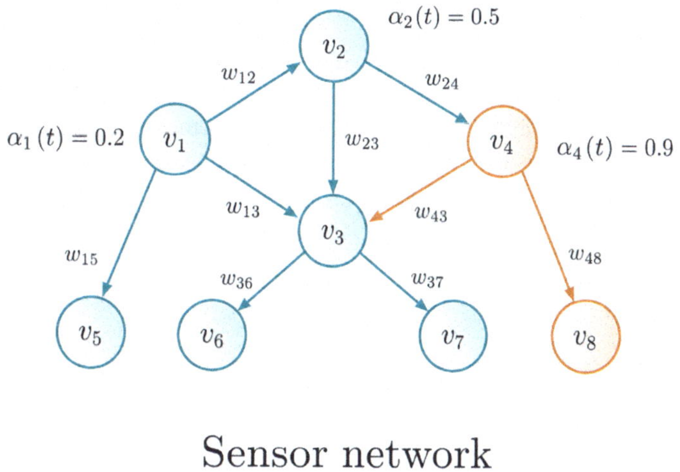

Fig. 3 Symbolic distributed adaptive sensor network

In this symbolic VO model:

- Rooftop sensors (e.g., v_1, v_2) have short-term memory caused by fast temperature or pollution variations.
- Underground sensors (e.g., v_4, v_9) have longer memory profiles, capturing slow hydrological or soil dynamics.

The symbolic VO-Laplacian facilitates context-based sensor memory adjustment, enhancing anomaly detection accuracy, energy-efficient communication, and network stability under fluctuating environmental conditions [15, 18, 22]. Building on these sensor-based applications, the framework can be further extended to symbolic simulation environments, enabling scalable experimentation and integration with hybrid AI systems.

6.3 *Symbolic Simulation Frameworks*

The symbolic VO-diffusion framework accommodates scalable symbolic simulation environments for:

- Analytical monitoring of node-wise memory development.
- Integration into symbol-based AI systems for adaptive decision-making in real time.

- Symbolic-numeric hybrid simulations that use memory dynamics alongside deep learning blocks.

Examples include symbolic memory-sensitive controllers in smart cities, cognitive robotic flocks, and fault-tolerant autonomous networks [23–25]. These promising avenues underscore the importance of advancing both the theoretical foundations and practical implementations of the symbolic VO-Laplacian, setting the stage for the following discussion on open challenges, limitations, and future research directions.

7 Discussion

The symbolic variable-order fractional Laplacian framework introduced here opens several avenues for research for theoretical formulation, numerical modeling, and practical applications.

7.1 Comparison with Fixed-Order Models

Fixed-order fractional models imply a homogeneous memory depth in all nodes and are unable to accommodate local adaptation and heterogeneity [1, 2]. The symbolic VO-fractional model, however, provides dynamic, node-dependent memory development, which better accommodates observed behavior in biological systems [5], social networks [17], and adaptive sensor grids [15].

7.2 Learning Memory Profiles from Data

In most applications, the node-specific memory profiles $\alpha_i(t)$ are not known in advance and have to be learned from observed dynamics. Machine learning models like recurrent neural networks (RNNs) [24] and memory-augmented graph neural networks (GNNs) [26] provide powerful tools for data-driven learning of changing memory patterns.

Hybrid symbolic models paired with data-driven memory learning would enable novel hybrid explainable AI systems to learn autonomously in changing environments [16, 18].

7.3 Symbolic Fractional Graph Neural Networks (Fractional-GNNs)

Graph neural networks have rapidly become the foundation of modern learning on structured data, with applications ranging from chemistry and biology to social networks and infrastructure systems. At their core, GNNs rely on the graph Laplacian or its variants to define message-transmission rules between nodes [27]. While effective, these conventional formulations inherit the limitations of fixed-order dynamics(i.e. all nodes are assumed to update information with the same temporal memory, regardless of context or history).

The symbolic VO-fractional Laplacian introduced in this chapter offers a natural pathway to overcome this restriction. By replacing the classical Laplacian with $L_{\alpha(t)}$, a GNN layer can propagate information through edges while simultaneously encoding heterogeneous and time-varying memory effects at each node. In such a *Fractional Graph Neural Network (F-GNN)*, the orders $\alpha_i(t)$ may be fixed, designed as functions of node states, or even learned directly during training. This flexibility aligns with recent efforts to incorporate fractional dynamics into graph learning and neural differential equations [19, 20, 24, 28–31]. It opens the door to neural architectures that not only process structure but also adaptively regulate how much of the past each node remembers.

Symbolic F-GNNs hold promise for a wide range of domains. In social dynamics, they could capture differences in user attention spans and memory of past interactions. In epidemiology, they may represent variable incubation or recovery memory in contagion networks. In transportation and traffic prediction, they can encode heterogeneous travel delays across nodes. Finally, in decentralized robotics or control systems, they could enable agents to adjust coordination rules based on memory depth.

Overall, symbolic fractional GNNs represent a compelling bridge between fractional calculus and machine learning, offering interpretability, adaptability, and resilience in learning on complex graphs. As research in this direction matures, they may form the basis for the next generation of memory-aware neural architectures.

7.4 Cognitive and Resilient Systems

The symbolic VO approach is relevant to with the emerging demands in cognitive networks [1], swarm robotics [32], and robust cyber-physical infrastructures [33]. Applications involving real-time adaptation, memory-based decision-making, and decentralized control can benefit from of symbolic VO-fractional modeling.

7.5 Multilayer and Temporal Networks

Generalizing the symbolic VO-fractional diffusion model to multilayer networks [23, 25] and temporally evolving graphs is a natural extension. Adaptation to interlayer or layer-specific memory can be included by assigning varying VO profiles to different network layers or interactions.

Memory evolution modeling over time and layers may offer strong insights into dynamic systems like financial markets [34], climate networks [35], and interconnected smart grids [36].

7.6 Fractional Control and Symbolic Optimization

Another promising direction is the construction of control methods based on symbolic VO-diffusion dynamics. Adaptive memory-variable control laws can be systematically designed to realize improved performance in tracking, stabilization, and disturbance rejection [37].

In addition, symbolic optimization structures based on VO memory dynamics may be formulated for energy-efficient routing, resource allocation, and adaptive communication in networked systems.

7.7 Limitations and Challenges

While the symbolic VO-Laplacian framework offers significant theoretical and practical advantages, several challenges remain. Estimating node-wise memory profiles $\alpha_i(t)$ from real-world data is often difficult and may require sophisticated inference or machine learning techniques. Secondly, the computational cost of simulating large-scale symbolic VO systems, particularly when combining symbolic and numerical elements, raises scalability concerns. Furthermore, integrating symbolic models into real-time control architectures and practical engineering systems introduces implementation hurdles that need to be carefully addressed. Overcoming these challenges will be essential for ensuring the framework's robustness, efficiency, and broad applicability.

8 Conclusion

A symbolic framework for heterogeneous memory-driven diffusion on graphs based on node-wise variable-order fractional calculus is presented in this chapter. Linearity and boundedness of $L_{\alpha(t)}$ ensure stable energy propagation in memory-based diffusion processes, and well-posedness guarantees that the system evolves deterministically under suitable conditions. These attributes are essentially analyzed for employing symbolic VO-diffusion models in adaptive systems such as sensor grids, cognitive networks, and decentralized learning environments. The chapter also demonstrates the VO symbolic simulation settings in the modeling flexibility and interpretability of the approach. Furthermore, the application of VO fractional symbolic model are discussed in learning memory profile from data, fractional GNN, cognitive and resilient system, multilayer temporal network and control and symbolic optimization.

Overall, the symbolic VO-fractional framework offers a scalable and transparent foundation for next-generation networked systems and memory-aware neural architectures, with promising implications across network science, artificial intelligence, and control.

References

1. Sun, H., Zhang, Y., Baleanu, D., Chen, W., Chen, Y.: A new collection of real-world applications of fractional calculus in science and engineering. Commun. Nonlinear Sci. Numer. Simul. **64**, 213–231 (2018)
2. Garrappa, R.: Numerical methods for fractional differential equations: a survey and a software tutorial. Mathematics **6**(2), 576 (2018)
3. Mainardi, F.: Fractional calculus and waves in linear viscoelasticity: an introduction to mathematical models. Imperial College Press, London (2010)
4. Metzler, R., Klafter, J.: The random walk's guide to anomalous diffusion: a fractional dynamics approach. Phys. Rep. **339**(1), 1–77 (2000)
5. Magin, R.L.: Fractional calculus in bioengineering. Begell House Publishers, Redding (2006)
6. Sahoo, A.K., Gupta, A.K.: Exploring solutions to the fractional Boiti-Leon-Manna-Pempinelli equation characterizing wave dynamics in incompressible fluids. Partial Differ. Equ. Appl. Math. **11**, 100903 (2024)
7. Saha Ray, S., Gupta, A.K.: A two-dimensional Haar wavelet approach for the numerical simulations of time and space fractional Fokker-Planck equations in modeling of anomalous diffusion systems. J. Math. Chem. **52**, 2277–2293 (2014)
8. Gupta, A.K., Saha Ray, S.: On the solutions of fractional Burgers-Fisher and generalized Fisher's equations using two reliable methods. Int. J. Math. Math. Sci. **2014**, 682910 (2014)
9. Gupta, A.K., Saha Ray, S.: An investigation with Hermite wavelets for accurate solution of fractional Jaulent-Miodek equation associated with energy-dependent Schrödinger potential. Appl. Math. Comput. **270**, 458–471 (2015)
10. Gupta, A.K., Saha Ray, S.: On the solitary wave solution of fractional Kudryashov-Sinelshchikov equation describing nonlinear wave processes in a liquid containing gas bubbles. Appl. Math. Comput. **298**, 1–12 (2017)
11. Nayak, S., Gupta, A.K.: Wavelet solutions for the time-fractional Phi-Four (ϕ^4) equation with Caputo derivatives. In: Kaur, L., Kumar, P. (eds.) Recent developments in fractional calculus: theory, applications, and numerical simulations, Studies in Systems, Decision and Control 235. Springer, Cham (2025)
12. Almeida, R., Pooseh, S., Torres, D.F.M.: A survey of variable-order fractional derivatives: mathematical foundations, applications and numerical methods. Appl. Math. Comput. **338**, 3–22 (2018)
13. Chung, F.: Spectral graph theory. CBMS Regional Conference Series in Mathematics, American Mathematical Society (1997)
14. Newman, M.E.J.: Networks: an introduction. Oxford University Press, Oxford (2010)

15. Elul, Y., Rozenberg, E., Bronstein, A.M.: Data-driven modeling of interrelated dynamical systems. Commun. Phys. **7**(1), 123 (2024)
16. Wang, Y., Hou, W., Sheng, N., et al.: Graph pooling in graph neural networks: methods and their applications in omics studies. Artif. Intell. Rev. **57**, 294 (2024)
17. Chen, P., Qi, M., Yan, L., Duan, X.: Diffusion capacity analysis of complex network based on the cluster distribution. Chaos Solitons Fract. **178**, 114329 (2024)
18. Tian, Y., Li, G., Sun, P.: Information evolution in complex networks. Chaos **32**(7), 073105 (2022)
19. Kang, K., Wang, H., Chen, Y., Li, Z.: Unleashing the potential of fractional calculus in graph neural networks with FROND. OpenReview , (2024)
20. Liu, X., Zhang, J., Yang, M., and Chen, Y.: Neural variable-order fractional differential equation. In: Proceedings of the AAAI Conference on Artificial Intelligence **39**(8), 12345–12353 (2025)
21. Kilbas, A.A., Srivastava, H.M., Trujillo, J.J.: Theory and applications of fractional differential equations. Elsevier, Amsterdam (2006)
22. Sibatov, R.T., L'vov, P.E., Sun, H.G.: Variable-order fractional diffusion: physical interpretation and simulation within the multiple trapping model. Appl. Math. Comput. **482**, 128960 (2024)
23. De Domenico, M., Granell, C., Porter, M.A., Arenas, A.: The physics of multilayer networks. Nat. Phys. **12**(10), 901–906 (2016)
24. Kang, Q., Zhao, K., Song, Y., Xie, Y., Zhao, Y., Wang, S., She, R., and Tay, W.P.: Coupling graph neural networks with non-integer order dynamics: a robustness study. NeurIPS 2023 Workshop: New Frontiers in Graph Learning (2023)
25. Boccaletti, S., Bianconi, G., Criado, R., del Genio, C.I., Gómez-Gardeñes, J., Romance, M., Sendiña-Nadal, I., Wang, Z., Zanin, M.: The structure and dynamics of multilayer networks. Phys. Rep. **544**(1), 1–122 (2014)
26. Ma, G., Vo, V.A., Willke, T.L., and Ahmed, N.K.: Memory-augmented graph neural networks: a neuroscience perspective. arXiv preprint arXiv:2209.10818 (2022)
27. Kipf, T.N., Welling, M.: Semi-supervised classification with graph convolutional networks. In: Proc. ICLR, (2017)
28. Kumar, P., Lee, T.H., Erturk, V.S.: A fractional-order multi-delayed bicyclic crossed neural network: stability, bifurcation, and numerical solution. Neural Netw. , 107436 (2025)
29. Kumar, P., Lee, T.H., Erturk, V.S.: A novel fractional-order neutral-type two-delayed neural network: stability, bifurcation, and numerical solution. Math. Comput. Simul. **232**, 245–260 (2025)
30. Kumar, P., Assali, E.A.: Fixed-time synchronization of fractional-order Hopfield neural networks with proportional delays. Math. Comput. Simul. **240**, 367–380 (2026)
31. Kumar, P., Lee, T.H., Erturk, V.S.: A novel two-delayed tri-neuron neural network with an incomplete connection. Nonlinear Dyn. **112**, 20269–20293 (2024)
32. Beni, G.: From swarm intelligence to swarm robotics. In: Swarm robotics, Lecture Notes in Computer Science 3342, Springer, Berlin, pp. 1–9 (2005)
33. Weerakkody, S., Ozel, O., Mo, Y., Sinopoli, B.: Resilient control in cyber-physical systems: countering uncertainty, constraints, and adversarial behavior. Found. Trends Syst. Control **7**(1–2), 1–252 (2019)
34. Battiston, S., Caldarelli, G., May, R.M., Roukny, T., Stiglitz, J.E.: The price of complexity in financial networks. Proc. Natl. Acad. Sci. U.S.A. **113**(36), 10031–10036 (2016)
35. Donges, J.F., Zou, Y., Marwan, N., Kurths, J.: The backbone of the climate network. EPL **87**(4), 48007 (2009)
36. Alilou, M., Azami, H., Oshnoei, A., Mohammadi-Ivatloo, B., Teodorescu, R.: Fractional-order control techniques for renewable energy and energy-storage-integrated power systems: a review. Fractal Fract. **7**(5), 391 (2023)
37. Ortigueira, M.D., Valério, D., Machado, J.T.: Variable order fractional systems. Commun. Nonlinear Sci. Numer. Simul. **71**, 231–243 (2019)

Existence of a Unique Solution of Fractional-Order Neural Network with Proportional Delay

Swati Tyagi

Abstract In this chapter, we consider neural network of fractional-order with order α and proportional time delay. We derive sufficient conditions for the existence of solutions of the delayed network for various initial conditions using fixed point theory and fractional calculus. Such models are considered suitable when a process takes place through strongly anomalous media. At the end, we give some numerical examples to show effectiveness of our analytic findings.

1 Introduction

The generalization of integer order differentiation and integral to non-integer order, also known as Fractional calculus has been introduced 300 years ago. For several years, the theory of fractional calculus was developed as a pure theoretical field and hence has been used by mathematicians only. However, in past few decades, the applications of fractional calculus to various fields such as biology, physics and engineering etc have attracted attention of researchers from other areas. This generalization of several dynamical equations using fractional calculus has been proved to be useful in modeling various real-world phenomena mathematically that arise from various interdisciplinary areas, such as control theory, viscoelasticity, signal processing and social sciences etc. For instance, we refer to [1–7]. Recently, the field of fractional calculus has emerged in the study of neural networks also. Delayed cellular neural networks (DCNNs) have been widely investigated and have found many important applications in the fields of pattern recognition, signal processing, optimization and associative memories, detecting speed of moving objects. To the best of the author's knowledge, few authors have considered dynamical behavior for the neural networks with proportional delays [8, 9], including global dissipativity of a class of cellular neural networks (CNNs) with multi-proportional delays. The study of exponential stability of a class of CNNs with multi-proportional delays,

S. Tyagi (✉)
Amity University Punjab, Mohali, India
e-mail: styagi@pb.amity.edu

L. Kaur and P. Kumar (eds.), *Neural Networks: Theory, Algorithms, Simulation, and Applications*, Studies in Computational Intelligence 1273,
https://doi.org/10.1007/978-3-032-18750-5_2

delay-independent stability criterion had been obtained by nonlinear measure in [9]. For more details on various results on fractional-order neural networks with and without time delay, we refer to [10–15] and references cited therein. However, one of the many delays, the proportional delay is considerately existent. The proportional delay is time-varying, unbounded in nature, and different from other types of delay (such as constant delay, bounded time-varying delay, distributed time delay etc.). Moreover, due to presence of amount of parallel pathways of a variety of axon sizes and lengths, neural networks possess spatial nature. Thus it becomes desirable to introduce continuous proportional delays over a certain duration of time. Analyzing dynamic behaviors of neural networks with proportional delays has gathered a considerable attention in past few years, such as global exponential stability, asymptotic stability and synchronization etc. For more details, we refer to [16–20]. To the best of our knowledge, the existence and uniqueness of solutions to fractional-order neural network with proportional delays is still challenging and open problem. For convenience, we consider Caputo fractional-order derivative in this chapter.

Motivated by the above, in this chapter, we consider the following fractional-order neural network model with proportional delay described by following equation:

$$ {}_C D_t^{\alpha} x_i(t) = -d_i x_i(t) + \sum_{j=1}^{n} a_{ij} f_j(x_j(t)) + \sum_{j=1}^{n} b_{ij} f_j(x_j(q_{ij}t)) + I_i \tag{1} $$

$$ x_i(0) = x_{i0}, \quad i = 1, 2, \cdots, n, \quad 0 < \alpha < 1, \tag{2} $$

where $n \geq 2$ represents the number of neurons in the neural network; ${}_C D_t^{\alpha}$ denotes the Caputo fractional derivative of fractional-order α; $x(t) = (x_1(t), x_2(t), \cdots, x_n(t))^T \in \mathbb{R}^n$ corresponds to the state vector at time t; a_{ij} and b_{ij} are the constants and denote the strength of connectivity between neuron j and i at time t and $q_{ij}t$ respectively; $q_{ij}t$, $i = 1, 2, \cdots, n$ are proportional delay factors satisfying $0 < q \leq 1$, $q = \min_{1 \leq i,j \leq n}\{q_{ij}\}$ and $f_i(\cdot)$ denotes non-linear activation function; I_i is the external bias and is a constant input.

This paper is organized as follows. In Sect. 2, we give some preliminary results of fractional calculus. In Sect. 3, we derive existence of solutions for the fractional-order neural network for various initial conditions using fixed point theory and fractional calculus. In Sect. 4, we give some examples to validate the obtained results. Finally we conclude our work in Sect. 5.

2 Preliminaries

In this section, we recall some definitions of fractional calculus, which are used later to derive main results. Let X denotes Banach space and $C(J, X)$ denotes Banach space of continuous functions $x(t)$ with $x(t) \in X$ for $t \in J := [0, T]$. Moreover, let $B_r(x, X)$ denotes the closed ball in X with center x and radius r. We denote $\|x\| = \max_{t\in J}$.

Definition 1 **The Euler Gamma Function** $\Gamma(z)$ is defined by the following integral:

$$\Gamma(z) = \int_0^{\infty} t^{z-1}e^{-t}dt, \quad Re(z) > 0, \quad z \in \mathbb{C}.$$

Definition 2 **The Fractional Integral** $D_{0,t}^{-\alpha}$ with fractional-order $\alpha > 0$ of a function $\zeta(t)$ is defined as

$$D_{0,t}^{-\alpha}\zeta(t) = \frac{1}{\Gamma(\alpha)} \int_0^t (t-s)^{\alpha-1}\zeta(s)ds, \quad t \geq 0.$$

Definition 3 **The Caputo derivative** with order $\alpha > 0$ of a function $\zeta(t) \in C^{n+1}([0, +\infty), R)$ is defined as

$${}_C D_{0,t}^{\alpha}\zeta(t) = \frac{1}{\Gamma(n-\alpha)} \int_0^t (t-s)^{n-\alpha-1}\zeta^{(n)}(s)ds, \quad t \geq 0, \quad n-1 < \alpha < n,$$

where n is a positive integer number.

Before proceeding, we assume that the following assumption holds true for the functions $f_i(\cdot)$, $i = 1, 2, \cdots, n$.

A1: For some constant $L_i > 0, i = 1, 2, \cdots, n$, each activation function $f_i(\cdot) : \mathbb{R} \to \mathbb{R}$ satisfies

$$|f_i(u) - f_i(v)| \leq L_{f_i}|u - v|, \quad \text{for} \quad u, v \in \mathbb{R}$$

We can write the model system (1) as the following equivalent integral equation:

$$x_i(t) = x_{i0} + \frac{1}{\Gamma(\alpha)} \int_0^t (t-s)^{\alpha-1}\Big[-d_i x_i(s) + \sum_{j=1}^{n} a_{ij} f_j(x_j(s)) + \sum_{j=1}^{n} b_{ij} f_j(x_j(q_{ij}s)) + I_i\Big]ds.$$

Denoting $x(t) = (x_1(t), x_2(t), \cdots, x_n(t))^T$, $d = diag(d_i)$, $A = (a_{ij})$, $B = (b_{ij})$, and $I = (I_1, I_2, \cdots, I_n)^T$, we can write the above system in the following matricial form as:

$$x(t) = x_0 + \frac{1}{\Gamma(\alpha)} \int_0^t (t-s)^{\alpha-1} [-dx(s) + Af(x(s)) + Bf(x(qs)) + I] ds. \quad (3)$$

For simplicity, we define

$$\beta = \frac{T^\alpha}{\Gamma(\alpha+1)}, \quad \text{and} \quad M = \max_{t\in[0,T]} \|f(0)\|. \tag{4}$$

3 Existence and Uniqueness of Solutions

Theorem 1 *If Assumption (A1) holds and $2\beta\xi < 1$, then the fractional-order neural network (3) with proportional delay has a unique solution.*

Proof Let $\Delta = C(J, X)$. We define a mapping $\Theta : \Delta \to \Delta$ by

$$\Theta x(t) = x_0 + \frac{1}{\Gamma(\alpha)} \int_0^t (t-s)^{\alpha-1} \left[-dx(s) + Af(x(s)) + Bf(x(qs)) + I\right] ds. \tag{5}$$

Then we have

$$\begin{aligned}
\|\Theta x(t)\| &\le \|x_0\| + \frac{d}{\Gamma(\alpha)} \int_0^t (t-s)^{\alpha-1} \|x(s)\| ds + \frac{\|A\|}{\Gamma(\alpha)} \int_0^t (t-s)^{\alpha-1} \|f(x(s))\| ds \\
&\quad + \frac{\|B\|}{\Gamma(\alpha)} \int_0^t (t-s)^{\alpha-1} \|f(x(qs))\| ds + \frac{\|I\|}{\Gamma(\alpha)} \int_0^t (t-s)^{\alpha-1} ds, \\
&\le \|x_0\| + \frac{d}{\Gamma(\alpha)} \int_0^t (t-s)^{\alpha-1} \|x(s)\| ds + \frac{\|A\|}{\Gamma(\alpha)} \int_0^t (t-s)^{\alpha-1} \|f(x(s)) - f(0)\| ds \\
&\quad + \frac{\|B\|}{\Gamma(\alpha)} \int_0^t (t-s)^{\alpha-1} \|f(x(qs)) - f(0)\| ds + \frac{\|I\|}{\Gamma(\alpha)} \int_0^t (t-s)^{\alpha-1} ds \\
&\quad + \frac{(\|A\| + \|B\|)}{\Gamma(\alpha)} \int_0^t (t-s)^{\alpha-1} \|f(0)\| ds, \\
&\le \|x_0\| + \frac{d}{\Gamma(\alpha+1)} T^\alpha r + \frac{(\|A\| + \|B\|) L_f}{\Gamma(\alpha+1)} T^\alpha r + \frac{((\|A\| + \|B\|) M + \|I\|)}{\Gamma(\alpha+1)} T^\alpha.
\end{aligned}$$

Letting $\dfrac{T^\alpha}{\Gamma(\alpha+1)} = \beta$, we obtain

$$\|\Theta x(t)\| = \|x_0\| + \xi\beta r + \gamma\beta,$$

where $\xi = d + (\|A\| + \|B\|) L_f$ and $\gamma = (\|A\| + \|B\|) M + \|I\|$. Choosing $r \ge 2(\|x_0\| + \gamma\beta)$, we obtain $\|\Theta x(t)\| \le r$. Thus $\Theta B_r \subset B_r$, where $B_r := \{x \in \Delta : \|x\| \le r\}$.

Furthermore, we have

$$\begin{aligned}\|\Theta x_1(t)-\Theta x_2(t)\| &\le \frac{1}{\Gamma(\alpha)}\int_0^t (t-s)^{\alpha-1}\Big[d\|x_1(s)-x_2(s)\|+\|A\|\|f(x_1(s))-f(x_2(s))\|\\ &\quad+\|B\|\|f(x_1(qs))-f(x_2(qs))\|ds\Big]\\ &\le \frac{T^\alpha}{\Gamma(\alpha+1)}\left[d+(\|A\|+\|B\|)L_f\right]\|x_1(t)-x_2(t)\|\\ &\le \xi\beta\|x_1(t)-x_2(t)\|.\end{aligned}$$

Since $2\xi\beta<1$, the mapping Θ is a contraction mapping and thus there exists a unique fixed point $x\in B_r$ such that $\Theta x(t)=x(t)$. □

Model with Non-local Initial Conditions: Now we derive the existence of solution for non-local initial conditions of the form

$$x(0)+g(x)=x_0, \tag{6}$$

where $g:\Delta\to X$ is a given function, which satisfies the following condition:

A2 g is Lipschitz continuous function, that is there exists a constant $G>0$, such that

$$\|g(x)-g(y)\|\le G\|x-y\|, \quad x,y,\in C(J,X).$$

Theorem 2 *Assume that Hypothesis (A1), (A2) hold. Furthermore,* $2(G+\beta\xi)<1$, *then the fractional-order neural network (3) with non-local initial condition (6) admits a unique solution.*

Proof We define operator $\Phi:\Delta\to\Delta$ by

$$\Phi x(t)=x_0-g(x)+\frac{1}{\Gamma(\alpha)}\int_0^t (t-s)^{\alpha-1}\left[-dx(s)+Af(x(s))+Bf(x(qs))+I\right]ds.$$

We now show that Φ has a fixed point, which is further a solution of system (1) and (6). Now

$$\begin{aligned}\|\Phi x(t)\| &\le \|x_0\|+\|g(x)-g(0)\|+\|g(0)\|+\frac{d}{\Gamma(\alpha)}\int_0^t (t-s)^{\alpha-1}\|x(s)\|ds\\ &\quad+\frac{\|A\|}{\Gamma(\alpha)}\int_0^t (t-s)^{\alpha-1}L_f\|x(s)\|ds+\frac{\|B\|}{\Gamma(\alpha)}\int_0^t (t-s)^{\alpha-1}L_f\|x(qs)\|ds\\ &\quad+\frac{\|A\|+\|B\|}{\Gamma(\alpha)}\int_0^t (t-s)^{\alpha-1}\|f(0)\|ds+\frac{\|I\|}{\Gamma(\alpha)}\int_0^t (t-s)^{\alpha-1}ds\\ &\le \|x_0\|+Gr+\|g(0)\|+dr\beta+(\|A\|+\|B\|)L_f r\beta+(\|A\|+\|B\|)M\beta\\ &\quad+\|I\|\beta\\ &\le \|x_0\|+Gr+\|g(0)\|+\xi r\beta+\gamma\beta.\end{aligned}$$

Choose $r \geq 2\left(\|x_0\| + \|g(0)\| + \gamma\beta\right)$. Then we have $\|\Phi x(t)\| \leq r$. Thus $\Phi B_r \subset B_r$. Further for $x_1, x_2 \in \Delta$, we have

$$\begin{aligned}\|\Phi x_1(t) - \Phi x_2(t)\| &\leq \|g(x_1) - g(x_2)\| + \frac{1}{\Gamma(\alpha)}\int_0^t (t-s)^{\alpha-1}\Big[d\|x_1(s) - x_2(s)\| \\ &\quad + \|A\|\|f(x_1(s)) - f(x_2(s))\| + \|B\|\|f(x_1(qs)) - f(x_2(qs))\|\Big]ds \\ &\leq \left(G + \left(d + (\|A\| + \|B\|)\, L_f\right)\beta\right)\|x_1(t) - x_2(t)\| \\ &= (G + \beta\xi)\,\|x_1(t) - x_2(t)\| \\ &\leq \frac{1}{2}\|x_1(t) - x_2(t)\|.\end{aligned}$$

Using contraction mapping principle, we obtain that Φ is a contraction mapping and thus there exists a unique fixed point $x \in B_r$, such that $\Phi x(t) = x(t)$. □

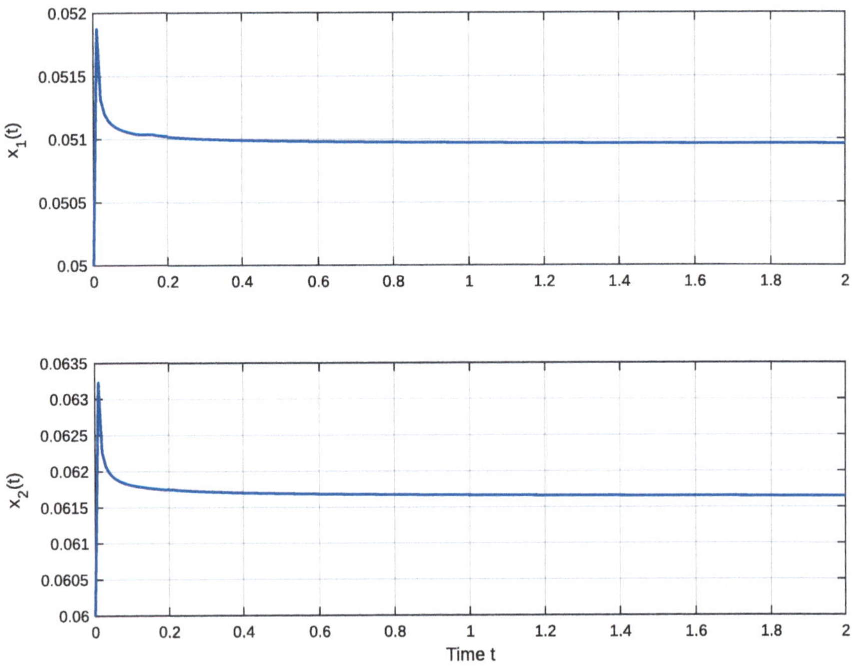

Fig. 1 Behaviour of solutions for $\alpha = 0.6$

4 Numerical Examples

Consider the following fractional-order neural network model with proportional delay:

$$ {}_C D^{\alpha} x(t) = -dx(t) + Af(x(s)) + Bf(x(qs)) + I, \qquad (7) $$

with
$d = \mathrm{diag}(0.02, 0.01)$, $A = \begin{pmatrix} 0.01 & 0.01 \\ 0 & 0.02 \end{pmatrix}$, $B = \begin{pmatrix} 0.04 & 0.02 \\ 0.01 & 0.03 \end{pmatrix}$, $I = (0.01, 0.02)^T$, $0 < q < 1$ and fractional-order $\alpha = 0.6$. The initial condition associated with (7) is $x_0 = (0.5, 0.6)^T$ and the nonlinear activation are chosen as $f(\cdot) = \tanh(\cdot)$. Then $2(\|x_0\| + \gamma\beta) = 0.12$. We can suitably choose $r = 1 > 0.12$, which implies $\|x\| \leq 1$. Moreover for suitable choice of $T = 10$, we obtain $\xi = 0.10$, $\beta = 4.4527$ and $\gamma = 0.10$. Thus we have $2\xi\beta = 0.89054 < 1$. Using Theorem 1, the fractional-order neural network (7) with proportional delay admits a unique solution. The corresponding simulations are shown in Fig. 1 to support the theoretical findings.

For $\alpha = 0.2$, we have $2(\|x_0\| + \gamma\beta) = 1.4762$. We can suitably choose $r = 2 > 1.4762$, which implies $\|x\| \leq 1$. Moreover for suitable choice of $T = 10$, we obtain $\xi = 0.08$, $\beta = 1.7261$ and $\gamma = 0.087$. Thus we have $2\xi\beta = 0.2762 < 1$, implying

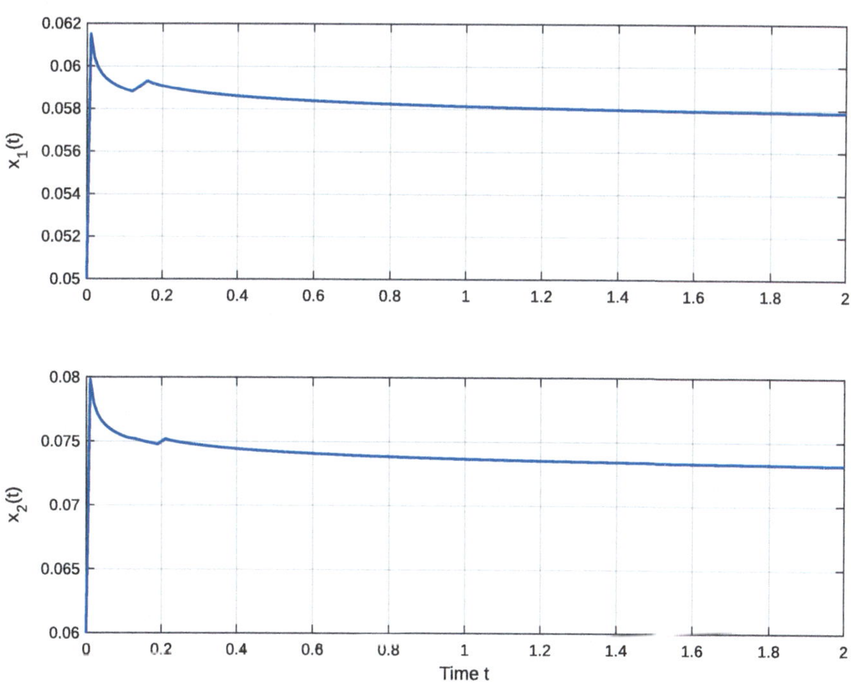

Fig. 2 Behaviour of solutions for $\alpha = 0.2$

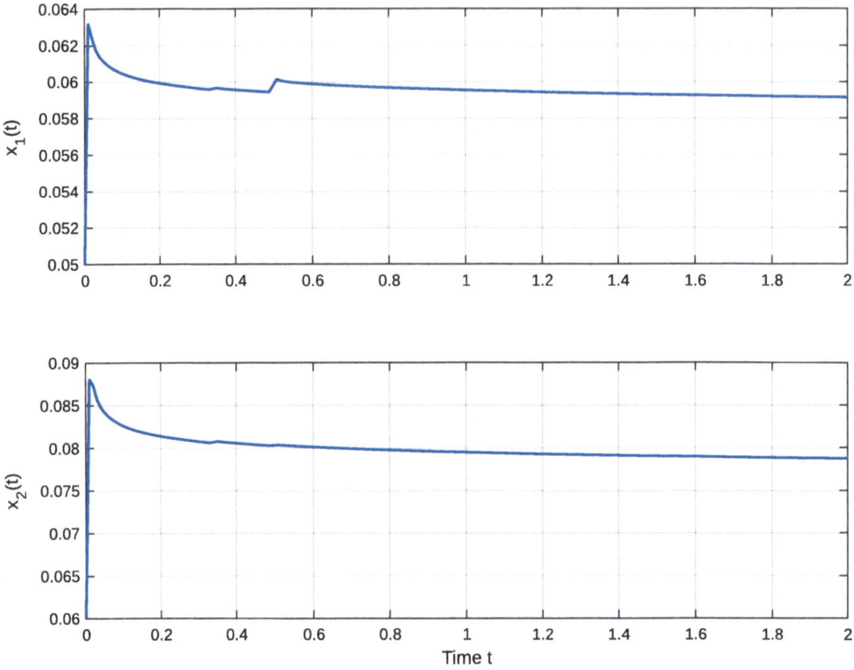

Fig. 3 Behaviour of solutions for $\alpha = 0.2$

the existence of a unique solution. The behaviour for fractional-order value $\alpha = 0.2$ is depicted in Fig. 2 below.

Furthermore, we check the solution behaviour for another set of parameter values considering

$$d = \text{diag}(0.01, 0.04),\ A = \begin{pmatrix} 0.02 & 0.04 \\ 0.03 & 0.20 \end{pmatrix},\ B = \begin{pmatrix} 0.01 & 0.03 \\ 0.01 & 0.01 \end{pmatrix},\ I = (0.01, 0.02)^T,$$

$0 < q < 1$ and fractional-order $\alpha = 0.2$. The initial condition associated with (7) is $x_0 = (0.5, 0.6)^T$ and the nonlinear activation are chosen as $f(\cdot) = \tanh(\cdot)$. Then $2(\|x_0\| + \gamma\beta) = 0.12$. We can suitably choose $r = 3 > 2.06305$, which implies $\|x\| \leq 1$. Moreover for suitable choice of $T = 10$, we obtain $\xi = 0.63$, $\beta = 1.7261$ and $\gamma = 0.86305$. Thus we have $2\xi\beta = 0.9321 < 1$. Using Theorem 1, the fractional-order neural network (7) with proportional delay admits a unique solution. The corresponding simulation results are shown in Fig. 3.

5 Conclusion

In this chapter, we have considered a fractional-order neural network with proportional delay. We have derived several sufficient conditions for the existence and uniqueness of solution to the considered model. The results are obtained in terms of simple algebric inequalities, which are less conservative and easy to verify. Our results improve the existing results obtained in the literature, which have been verified by numerical examples.

References

1. Caputo, M., Mainardi, F.: A new dissipation model based on memory mechanism. Pure Appl. Geophys. **91**, 134–147 (1971)
2. Mainardi, F., Raberto, M., Gorenflo, R., Scalas, E.: Fractional calculus and continuous-time finance II: the waiting-time distribution. XXPhys. A **287**, 468–481 (2000)
3. Celik, V., Demir, Y.: Chaotic fractional order delayed cellular neural network. In New trends in nanotechnology and fractional calculus applications 313–320, Springer Netherlands (2010)
4. Stamova, I.: Global Mittag-Leffler stability and synchronization of impulsive fractional-order neural networks with time-varying delays. Nonlinear Dyn. **77**, 1251–1260 (2014)
5. Hirsch, M.W.: Convergent activation dynamics in continuous time networks. Neural Netw. **2**, 331–349 (1989)
6. Delavari, H., Baleanu, D., Sadati, J.: Stability analysis of Caputo fractional-order nonlinear systems revisited. Nonlinear Dyn. **67**, 2433–2439 (2012)
7. Guo, S., Huang, L.: Stability of nonlinear waves in a ring of neurons with delays. J. Differential Equations **236**, 343–374 (2007)
8. Zhou, L.: On the global dissipativity of a class of cellular neural networks with multipantograph delays. Adv Artif Neural Syst , 941426 (2011)
9. Zhang, Y., Zhou, L.: Exponential stability of a class of cellular neural networks with multi-pantograph delays. Acta Electronica Sinica (China) **40**, 1159–1163 (2012)
10. Chen, L., Chai, Y., Wu, R., Ma, T., Zhai, H.: Dynamic analysis of a class of fractional-order neural networks with delay. Neurocomputing **111**, 190–194 (2013)
11. Wang, H., Yu, Y., Wen, G.: Stability analysis of fractional-order Hopfield neural networks with time delays. Neural Netw. **55**, 98–109 (2014)
12. Chen, H., Zhong, S., Yang, J.: A new globally exponential stability criterion for neural networks with discrete and distributed delays. Math. Probl. Eng. , 1–9 (2015)
13. Ran-Chao, W., Xin-Dong, H., Li-Ping, C.: Finite-time stability of fractional-order neural networks with delay. Commun. Theor. Phys. **60**, 189–193 (2013)
14. Tyagi, S., Abbas, S., Hafayed, M.: Global Mittag-Leffler stability of complex valued fractional-order neural network with discrete and distributed delays. Rendiconti del Circolo Matematico di Palermo Series **2**(65), 485–505 (2016)
15. Tyagi, S., Abbas, S.: Stability and synchronization of delayed fractional-order projection neural network with piecewise constant argument of mixed type. Tbilisi Mathematical Journal **10**, 57–74 (2017)
16. Balachandran, K., Kiruthika, S., Trujillo, J.J.: Existence of solutions of nonlinear fractional pantograph equations. Acta Mathematica Scientia **33**, 712–720 (2013)
17. Iserles, A., Liu, Y.: On neutral functional-differential equations with proportional delays. J. Math. Anal. Appl. **207**, 73–95 (1997)
18. Liu, Y.: Asymptotic behaviour of functional-differential equations with proportional time delays. Eur. J. Appl. Math. **7**, 11–30 (1996)
19. Kumar, P., Assali, E.A.: Fixed-time synchronization of fractional-order Hopfield neural networks with proportional delays. Math. Comput. Simul. , (2025)
20. Kumar, P., Lee, T.H., Erturk, V.S.: A fractional-order multi-delayed bicyclic crossed neural network: Stability, bifurcation, and numerical solution. Neural Netw. **188**, 107436 (2025)

Synchronization of Inertial Complex-Valued Competitive Neural Networks with Proportional Delays via Non-reduced Order Method

El Abed Assali and Ruiqi Li

Abstract Control and synchronization are very important aspects of any dynamic system. In this chapter, we investigate the exponential synchronization of a widely studied type of artificial neural network, namely competitive neural networks with an inertial term and proportional delays, in the complex domain $\mathbb{C}$. Unlike traditional techniques, a non-separation and non-reduced-order approach is employed to derive new and practical criteria for the exponential synchronization of the considered model. The analysis is conducted through the construction of novel Lyapunov functions, and a numerical example with corresponding simulation results demonstrates the effectiveness of the proposed method.

1 Introduction

Neural networks have been widely used in many fields, and research has shown that they can handle a bunch of challenges related to optimization, communication security, associative memory, and other dynamic issues [1–7]. Among various types of artificial neural networks, competitive neural networks (CoNNs) have attracted significant attention and have been widely investigated by researchers ([8, 9, 11]). Note that, Meyer-Bäse et al. [12] first introduced ComptNNs, which involve two main types of state variables: short-term memory (STM) and long-term memory (LTM). The STM characterizes rapid neural processes, while the LTM corresponds to

E. A. Assali (✉)
University of Jendouba, Higher School of Engineering of Medjez el Bab (ESIM), Route du Kef Km 5, Medjez el Bab, Tunisia

University of Carthage, Faculty of Sciences of Bizerte, Department of Mathematics, Research Laboratory GAMA LR21ES10, 7021 Zarzouna, Bizerta, Tunisia
e-mail: elabed.assali@fsb.rnu.tn

R. Li
School of Electrical and Information Engineering, Hunan Institute of Technology, Hengyang 421002, China
e-mail: 17742522911@163.com

L. Kaur and P. Kumar (eds.), *Neural Networks: Theory, Algorithms, Simulation, and Applications*, Studies in Computational Intelligence 1273,
https://doi.org/10.1007/978-3-032-18750-5_3

slower, unsupervised synaptic modifications [9, 10]. Owing to this structure, CoNNs can effectively model the dynamics of cortical cognitive maps, making the study of their dynamic behavior particularly meaningful.

As commonly acknowledged, synchronisation is an example of an extremely important dynamic behaviour in neural networks, and has proven successful in diverse domains, including information science [13] and secure communication [2]. In recent times, various research results on the synchronisation of CoNNs have been published [14–18]. In [14], the fixed-time synchronization problem of CoNNs with different time scales was studied by using Lyapunov stability theory. In [15], several sufficient conditions for cluster synchronization of coupled CoNNs with delays and two time scales under an adaptive pinning controller is designed and a modified Lyapunov functional. By constructing a novel Lyapunov–Krasovskii functional, a delay-delay partitioning approach was designed to achieve the synchronization problem for a class of delayed CoNNs with different time scales [16]. In [17], adaptive lag synchronization for a kind of CoNNs with discrete and mixed delays was investigated by constructing a proper Lyapunov functional and using robust adaptive controller.

In 1986, a novel type of neural network, known as an inertial neural network, was proposed by incorporating an inductor into the neural circuit to introduce inertial properties. The dynamics of these networks are then modeled using second-order differential equations [19].

Inertial neural networks, which draw inspiration from both engineering physics and biology [20], have attracted significant attention in recent years. Research on these networks has produced valuable results in several areas, including the stability of equilibrium points [21], the stability of periodic or anti-periodic solutions [22, 23], pinning synchronization [24], and fixed-time synchronization [2]. A common approach in earlier studies has been to reduce inertial neural networks to first-order differential equations. While this reduced-order method is effective, it also introduces challenges. Specifically, transforming a second-order inertial neural network into a first-order system via variable substitution doubles the system size, thereby increasing the complexity and complicating theoretical analysis [25].

Recently, non-reduced order techniques have been applied to study synchronization [26], stability [27, 28], and anti-periodicity [29] in various inertial neural network models. However, these studies predominantly focus on real-valued networks, often overlooking the role of complex-valued dynamics that appear in practical applications.

In practical applications, complex-valued variables play a crucial role, such as in nonlinear filtering [30] and image reconstruction [31]. Compared to real-valued neural networks, complex-valued neural networks can tackle a wider range of problems. A well-known example is the XOR problem, which cannot be solved by real-valued networks but is easily addressed using complex-valued networks [32]. To date, numerous studies have focused on the synchronization and stability of first-order complex-valued CoNNs through the separation of complex-valued models into two distinct real-valued parts [33–35]. However, the non-separation approach has been shown to be more efficient and practical [36].

In practical systems, delays are inevitable due to the limited exchange of information between different units, which can lead to divergences, instability, or oscillations [37–39]. Hence, when studying the dynamics of complex-valued CoNNs, delays must be taken into account. Complex-valued CoNNs have been studied by some researchers, but most studies focus primarily on time-bound delays [34, 35] and mixed delays [33]. However, there are no other chapters that have examined the synchronization of inertial complex-valued competitive neural networks (InCV-CoNNs) with unbounded delays. In contrast to the types of delay mentioned above, proportional delay is a class of delay that is time varying and unbounded. Zhou first proposed proportional delay in neural networks in 2011 [40]. To our knowledge, no research has yet been focused on the exponential synchronization (ExpS) of InCV-CoNNs with proportional delays using a combination of the non-separation method and non-reduced order formulation.

Inspired by the above discussion, this chapter aims to study the ExpS of a class of delayed InCVCoNNs using a direct analysis method and non-separation approach. In this chapter, we first propose a class of InCVCoNNs with time-varying delays. Second, by developing new Lyapunov functions with no transformation of variables, which is totally unlike the conventional reduced-order method, we prove the ExpS of this class of neural networks.

The remainder of this chapter is organized as follows: Sect. 2 introduces the preliminaries and outlines the proposed model. Section 3 examines the ExpS of the model. Section 4 provides a numerical illustration, and Sect. 5 concludes the study.

Notations

In this chapter, the following notations are adopted. The imaginary unit is denoted by i, where $i^2 = -1$. The sets $\mathbb{R}$ and $\mathbb{C}$ represent the real and complex numbers, while $\mathbb{R}^n$ and $\mathbb{C}^n$ correspond to the n-dimensional real and complex vector spaces, respectively.

For any $\mathrm{w} \in \mathbb{C}$, $\Re(\mathrm{w})$ and $\Im(\mathrm{w})$ denote its real and imaginary parts. The modulus of w is expressed as

$$|\mathrm{w}| = \sqrt{\mathrm{w}\,\overline{\mathrm{w}}}.$$

For a vector $\mathrm{W} = (\mathrm{w}_1, \ldots, \mathrm{w}_n)^T$, the Euclidean norm is given by

$$\|\mathrm{W}\| = \sqrt{\sum_{i=1}^{n} |\mathrm{w}_i|^2}.$$

2 System Description and Preliminaries

In this chapter, we study the following InCVCoNNs with proportional delays. The general dynamic form of these networks is given by the following equation:

$$\begin{cases} \text{STM}: \mathrm{w}_i''(\mathrm{t}) + a_i \mathrm{w}_i'(\mathrm{t}) = -d_i \mathrm{w}_i(\mathrm{t}) + \sum_{j=1}^{n} E_{ij} f_j\left(\mathrm{w}_j(\mathrm{t})\right) + \sum_{j=1}^{n} H_{ij} f_j\left(\mathrm{w}_j(\kappa \mathrm{t})\right) \\ \qquad + c_i \sum_{l=1}^{q} T_l m_{il}(\mathrm{t}) + I_i, \\ \text{LTM}: m_{il}''(\mathrm{t}) + m_{il}'(\mathrm{t}) = -m_{il}(\mathrm{t}) + T_l f_i(\mathrm{w}_i(\mathrm{t})), \end{cases} \tag{1}$$

where $\mathrm{t} \geq \mathrm{t}_0 \geq 0$, $i = 1, 2, \ldots, n$ and $l = 1, 2, \ldots, q$. The parameters $a_i, d_i > 0$ are the time constants of the neurons. $\mathrm{w}_i(\cdot) \in \mathbb{C}$ denotes the current activity of the ith neuron, and $m_{il}(\cdot) \in \mathbb{C}$ represents the corresponding synaptic efficiency. In general, a proportional delay is characterized by a constant factor κ satisfying $0 < \kappa < 1$. The corresponding time relation is described by $\kappa \mathrm{t} = \mathrm{t} - (1 - \kappa)\mathrm{t}$, where $(1 - \kappa)\mathrm{t}$ represents the proportional delay associated with the current time t. The term T_l represents a constant external stimulus with strength c_i, while $E_{ij} \in \mathbb{C}$ denotes the connection weight between neurons i and j, and $H_{ij} \in \mathbb{C}$ indicates the synaptic weight associated with delayed feedback. Finally, I_i is a constant external input.

First, the system in (1) can be rewritten in the following form:

Let $s_i(\mathrm{t}) = \sum_{l=1}^{q} T_l m_{il}(\mathrm{t})$ denote the sum of the long-term memory (LTM) over l. Based on equation (1), InCVCoNNs with proportional delays can be expressed in the following dynamic form:

$$\begin{cases} \text{STM}: \mathrm{w}_i''(\mathrm{t}) + a_i \mathrm{w}_i'(\mathrm{t}) = -d_i \mathrm{w}_i(\mathrm{t}) + \sum_{j=1}^{n} E_{ij} f_j\left(\mathrm{w}_j(\mathrm{t})\right) + \sum_{j=1}^{n} H_{ij} f_j\left(\mathrm{w}_j(\kappa \mathrm{t})\right), \\ \qquad + c_i s_i(\mathrm{t}) + I_i, \\ \text{LTM}: s_i''(\mathrm{t}) + s_i'(\mathrm{t}) = -s_i(\mathrm{t}) + \|T\|^2 f_i(\mathrm{w}_i(\mathrm{t})), \end{cases} \tag{2}$$

where $\|T\|^2 = T_1^2 + \ldots + T_q^2$ is a constant. Without loss of generality, the input stimulus T is assumed to be normalized such that its magnitude satisfies $\|T\|^2 = 1$. Based on the preceding assumption, the InCVCoNNs model can be reduced to the following simplified system:

$$\begin{cases} \text{STM}: \mathrm{w}_i''(\mathrm{t}) + a_i \mathrm{w}_i'(\mathrm{t}) = -d_i \mathrm{w}_i(\mathrm{t}) + \sum_{j=1}^{n} E_{ij} f_j\left(\mathrm{w}_j(\mathrm{t})\right) + \sum_{j=1}^{n} H_{ij} f_j\left(\mathrm{w}_j(\kappa \mathrm{t})\right), \\ \qquad + c_i s_i(\mathrm{t}) + I_i, \\ \text{LTM}: s_i''(\mathrm{t}) + s_i'(\mathrm{t}) = -s_i(\mathrm{t}) + f_i(\mathrm{w}_i(\mathrm{t})), \end{cases} \tag{3}$$

The initial conditions of neural networks (3) are defined as follows:

$$\begin{cases} \mathrm{w}_i(s) = \phi_i(s),\ \mathrm{w}_i'(s) = \tilde{\phi}_i(s),\ s \in [\kappa \mathrm{t}_0, \mathrm{t}_0], \\ s_i(s) = \psi_i(s),\ s_i'(s) = \tilde{\psi}_i(s),\ s \in [\kappa \mathrm{t}_0, \mathrm{t}_0], \end{cases} \quad i = 1, 2, \ldots, n. \tag{4}$$

Assuming system (3) is selected as the drive system, then the response system can be defined as follows:

$$\begin{cases} \mathrm{STM}: v_i''(\mathrm{t}) + a_i v_i'(\mathrm{t}) = -d_i v_i(\mathrm{t}) + \sum_{j=1}^{n} E_{ij} f_j\left(v_j(\mathrm{t})\right) + \sum_{j=1}^{n} H_{ij} f_j\left(v_j(\kappa \mathrm{t})\right) \\ \qquad\qquad + c_i q_i(\mathrm{t}) + I_i + \mathcal{U}_i(\mathrm{t}), \\ \mathrm{LTM}: q_i''(\mathrm{t}) + q_i'(\mathrm{t}) \quad = -q_i(\mathrm{t}) + f_i(v_i(\mathrm{t})) + \mathcal{V}_i(\mathrm{t}), \end{cases} \tag{5}$$

where $v_i(\cdot) \in \mathbb{C}$ and $q_i(\cdot)$ denote the states of the ith neuron in the response system (5), and $\mathcal{U}_i(\cdot)$ and $\mathcal{V}_i(\cdot)$ represent the feedback controllers.

The initial values of system (5) are given as follows :

$$\begin{cases} v_i(s) = \widehat{\phi}_i(s),\ v_i'(s) = \widehat{\tilde{\phi}}_i(s),\ s \in [\kappa \mathrm{t}_0, \mathrm{t}_0], \\ q_i(s) = \widehat{\psi}_i(s),\ q_i'(s) = \widehat{\tilde{\psi}}_i(s),\ s \in [\kappa \mathrm{t}_0, \mathrm{t}_0], \end{cases} \quad i = 1, 2, \ldots, n. \tag{6}$$

Definition 1 The InCVCoNNs systems (3) and (5) are said to be globally exponentially synchronized if, for every solution $\mathrm{w}(\mathrm{t}) = (\mathrm{w}_1, \ldots, \mathrm{w}_n)$, $s(\mathrm{t}) = (s_1, \ldots, s_n)$, $v(\mathrm{t}) = (v_1, \ldots, v_n)$, and $q(\mathrm{t}) = (q_1, \ldots, q_n)$ that satisfies (4) and (6), there exist a positive real number $M > 0$ (which may depend on the initial conditions) and a constant $\epsilon > 0$ such that

$$\|v(\mathrm{t}) - \mathrm{w}(\mathrm{t})\| + \|q(\mathrm{t}) - s(\mathrm{t})\| \leq M e^{-\epsilon \mathrm{t}}, \quad \mathrm{t} \geq 0.$$

In the present chapter, the following assumption is made about the activation functions:

(H.1) For every $i = 1, 2, ..., n$, there exist constant $L_i^f > 0$, ensuring that, for all $\mathrm{w}_1,\ \mathrm{w}_2 \in \mathbb{C}$, the following condition is met:

$$|f_i(\mathrm{w}_1) - f_i(\mathrm{w}_2)| \leq L_i^f |\mathrm{w}_1 - \mathrm{w}_2|,$$

Let $e_i(\mathrm{t}) = v_i(\mathrm{t}) - \mathrm{w}_i(\mathrm{t})$ and $\theta_i(\mathrm{t}) = q_i(\mathrm{t}) - s_i(\mathrm{t})$ denote the synchronization errors for $i = 1, 2, \ldots, n$. Therefore, the corresponding error system can be easily expressed as

$$
\begin{cases}
\text{STM}: e_i''(t) + a_i e_i'(t) = -d_i e_i(t) + \sum_{j=1}^{n} E_{ij}\left(f_j\left(v_j(t)\right) - f_j\left(w_j(t)\right)\right), \\
\qquad\qquad + \sum_{j=1}^{n} H_{ij}\left(f_j\left(v_j(\kappa t)\right) - f_j\left(w_j(\kappa t)\right)\right) \\
\qquad\qquad + c_i\theta_i(t) + \mathcal{U}_i(t), \\
\text{LTM}: \theta_i''(t) + \theta_i'(t) \quad = -\theta_i(t) + \left(f_i(v_i(t)) - f_i(w_i(t))\right) + \mathcal{V}_i(t).
\end{cases}
\tag{7}
$$

The control laws $\mathcal{U}_i(\cdot)$ and $\mathcal{V}_i(\cdot)$ are introduced in the following form:

$$
\begin{cases}
\mathcal{U}_i(t) = -\alpha_i e_i(t) - \beta_i e_i'(t) - \sum_{j=1}^{n} H_{ij}\left(f_j\left(v_j(\kappa t)\right) - f_j\left(w_j(\kappa t)\right)\right), \\
\mathcal{V}_i(t) = -\widetilde{\alpha}_i\theta_i(t) - \widetilde{\beta}_i\theta_i'(t),
\end{cases}
\tag{8}
$$

where, $\alpha_i > 0,\ \beta_i > 0,\ \widetilde{\alpha}_i > 0$, and $\widetilde{\beta}_i > 0$ denote the control gains.

Substituting $\mathcal{U}_i(t)$ and $\mathcal{V}_i(t)$ as defined in (8), we get

$$
\begin{cases}
\text{STM}: e_i''(t) + a_i e_i'(t) = -d_i e_i(t) + \sum_{j=1}^{n} E_{ij}\left(f_j\left(v_j(t)\right) - f_j\left(w_j(t)\right)\right), \\
\qquad\qquad + c_i\theta_i(t) - \alpha_i e_i(t) - \beta_i e_i'(t), \\
\text{LTM}: \theta_i''(t) + \theta_i'(t) \quad = -\theta_i(t) + \left(f_i(v_i(t)) - f_i(w_i(t))\right) - \widetilde{\alpha}_i\theta_i(t) - \widetilde{\beta}_i\theta_i'(t).
\end{cases}
\tag{9}
$$

The following contributions are presented to obtain the synchronization results:

(H.2) For each $i = 1, \cdots, n$ there exist nonzero constants $\lambda_i, \mu_i, \nu_i, \sigma_i$ and positive numbers δ_i, η_i and ϵ such that

$$
\Gamma_i < 0,\ \ \widetilde{\Gamma}_i < 0,
$$

and

$$
\Lambda_i^2 < 4\Delta_i\Gamma_i,\ \ \widetilde{\Lambda}_i^2 < 4\widetilde{\Delta}_i\widetilde{\Gamma}_i,
$$

where

$$
\begin{cases}
\Gamma_i = \lambda_i^2(\epsilon - a_i - \beta_i) + \lambda_i\mu_i + \frac{1}{2}\sum_{j=1}^{n}\lambda_i^2 L_j^f |E_{ij}|, \\
\Delta_i = \epsilon\mu_i^2 + \epsilon\delta_i + \frac{1}{2}L_i^f\left(\nu_i^2 + \nu_i\sigma_i\right) - \lambda_i\mu_i\left(d_i + \alpha_i\right) \\
\quad + \frac{1}{2}\sum_{j=1}^{n}\left(L_i^f|E_{ji}|\left(\lambda_j^2 + \lambda_j\mu_j\right) + \lambda_i\mu_i L_j^f|E_{ij}|\right), \\
\Lambda_i = \delta_i + \lambda_i\mu_i(2\epsilon - a_i - \beta_i) + \mu_i^2 - \lambda_i^2\left(d_i + \alpha_i\right),
\end{cases}
$$

and

$$\begin{cases} \widetilde{\Gamma}_i = \nu_i^2 \left(\epsilon + \frac{1}{2}L_i^f - \widetilde{\beta}_i - 1\right) + \nu_i\sigma_i, \\ \widetilde{\Delta}_i = \epsilon\sigma_i^2 + \epsilon\eta_i + \frac{1}{2}\nu_i\sigma_i L_i^f - \nu_i\sigma_i - \nu_i\sigma_i\widetilde{\alpha}_i, \\ \widetilde{\Lambda}_i = 2\epsilon\nu_i\sigma_i + \eta_i - \nu_i^2 - \nu_i\sigma_i + \sigma_i^2\nu_i^2\widetilde{\alpha}_i - \nu_i\sigma_i\widetilde{\beta}_i. \end{cases}$$

3 Exponential Synchronization

This section derives several sufficient conditions that ensure the ExpS of the neural systems described by (3) and (5).

Theorem 1 *Using* ***(H.1)*** *and* ***(H.2)****, the ExpS of the InCVCoNNs system (3) and the InCVCoNNs system (5) is achieved by implementing the controller (8).*

Proof We can define the following Lyapunov functional

$$L(\mathrm{t}) = L_1(\mathrm{t}) + L_2(\mathrm{t}) + L_3(\mathrm{t}) + L_4(\mathrm{t}) \tag{10}$$

where

$$L_1(\mathrm{t}) = \frac{1}{2}\sum_{i=1}^{n} \delta_i e_i(\mathrm{t})\overline{e_i(\mathrm{t})}e^{2\epsilon\mathrm{t}} + \frac{1}{2}\sum_{i=1}^{n} \eta_i\theta_i(\mathrm{t})\overline{\theta_i(\mathrm{t})}e^{2\epsilon\mathrm{t}},$$

$$L_2(\mathrm{t}) = \frac{1}{2}\sum_{i=1}^{n} e^{2\epsilon\mathrm{t}}\left(\lambda_i e_i'(\mathrm{t}) + \mu_i e_i(\mathrm{t})\right)\overline{\left(\lambda_i e_i'(\mathrm{t}) + \mu_i e_i(\mathrm{t})\right)},$$

and

$$L_3(\mathrm{t}) = \frac{1}{2}\sum_{i=1}^{n} e^{2\epsilon\mathrm{t}}\left(\nu_i\theta_i'(\mathrm{t}) + \sigma_i\theta_i(\mathrm{t})\right)\overline{\left(\nu_i\theta_i'(\mathrm{t}) + \sigma_i\theta_i(\mathrm{t})\right)},$$

Taking derivative of $L_1(\mathrm{t})$, then we have

$$\begin{aligned} \frac{dL_1(\mathrm{t})}{dt} &= e^{2\epsilon\mathrm{t}}\sum_{i=1}^{n}\Bigg[\epsilon\delta_i e_i(\mathrm{t})\overline{e_i(\mathrm{t})} + \frac{1}{2}\delta_i\left(e_i'(\mathrm{t})\overline{e_i(\mathrm{t})} + e_i(\mathrm{t})\overline{e_i'(\mathrm{t})}\right) \\ &\quad + \epsilon\eta_i\theta_i(\mathrm{t})\overline{\theta_i(\mathrm{t})} + \frac{1}{2}\eta_i\left(\theta_i'(\mathrm{t})\overline{\theta_i(\mathrm{t})} + \theta_i(\mathrm{t})\overline{\theta_i'(\mathrm{t})}\right)\Bigg], \end{aligned} \tag{11}$$

Deriving the derivative of $L_2(\mathrm{t})$ along (9), we can obtain that

$$\begin{aligned}
\frac{dL_2(\mathrm{t})}{dt} &= e^{2\epsilon \mathrm{t}} \sum_{i=1}^{n} \Bigg[\epsilon \left(\lambda_i e_i'(\mathrm{t}) + \mu_i e_i(\mathrm{t})\right) \overline{\left(\lambda_i e_i'(\mathrm{t}) + \mu_i e_i(\mathrm{t})\right)} \\
&+ \frac{1}{2} \left(\lambda_i e_i''(\mathrm{t}) + \mu_i e_i'(\mathrm{t})\right) \overline{\left(\lambda_i e_i'(\mathrm{t}) + \mu_i e_i(\mathrm{t})\right)} \\
&+ \frac{1}{2} \left(\lambda_i e_i'(\mathrm{t}) + \mu_i e_i(\mathrm{t})\right) \overline{\left(\lambda_i e_i''(\mathrm{t}) + \mu_i e_i'(\mathrm{t})\right)} \Bigg] \\
&= e^{2\epsilon \mathrm{t}} \sum_{i=1}^{n} \Bigg[\left(\epsilon \lambda_i^2 - \lambda_i^2 a_i + \lambda_i \mu_i - \lambda_i^2 \beta_i\right) e_i'(\mathrm{t}) \overline{e_i'(\mathrm{t})} \\
&+ \left(\epsilon \mu_i^2 - \lambda_i \mu_i d_i - \lambda_i \mu_i \alpha_i\right) e_i(\mathrm{t}) \overline{e_i(\mathrm{t})} \\
&+ 2 \left(\epsilon \lambda_i \mu_i - \frac{1}{2} \lambda_i \mu_i a_i + \frac{1}{2} \mu_i^2 - \frac{1}{2} \lambda_i^2 d_i - \frac{1}{2} \lambda_i^2 \alpha_i - \frac{1}{2} \lambda_i \mu_i \beta_i\right) \Re(e_i'(\mathrm{t}) \overline{e_i(\mathrm{t})}) \\
&+ \lambda_i^2 \sum_{j=1}^{n} \Re\Big(E_{ij} \left(f_j\left(v_j(\mathrm{t})\right) - f_j\left(\mathrm{w}_j(\mathrm{t})\right)\right) \overline{e_i'(\mathrm{t})} \Big) \\
&+ \lambda_i \mu_i \sum_{j=1}^{n} \Re\Big(E_{ij} \left(f_j\left(v_j(\mathrm{t})\right) - f_j\left(\mathrm{w}_j(\mathrm{t})\right)\right) \overline{e_i(\mathrm{t})} \Big)
\end{aligned} \tag{12}$$

According to **(H.1)** and the basic inequality $|\mathrm{w}_1 \overline{\mathrm{w}}_2| \leq \frac{1}{2}\left(|\mathrm{w}_1|^2 + |\mathrm{w}_2|^2\right) \quad (\mathrm{w}_1,\ \mathrm{w}_2 \in \mathbb{C})$ that

$$\begin{aligned}
&\sum_{i=1}^{n} \sum_{j=1}^{n} \lambda_i^2 \Re\Big(E_{ij} \left(f_j\left(v_j(\mathrm{t})\right) - f_j\left(\mathrm{w}_j(\mathrm{t})\right)\right) \overline{e_i'(\mathrm{t})} \Big) \\
&\leq \sum_{i=1}^{n} \sum_{j=1}^{n} \lambda_i^2 \Big| E_{ij} \left(f_j\left(v_j(\mathrm{t})\right) - f_j\left(\mathrm{w}_j(\mathrm{t})\right)\right) \overline{e_i'(\mathrm{t})} \Big| \\
&\leq \sum_{i=1}^{n} \sum_{j=1}^{n} \lambda_i^2 L_j^f |E_{ij}| |\overline{e_i'(\mathrm{t})}| |e_j(\mathrm{t})| \\
&\leq \frac{1}{2} \sum_{i=1}^{n} \sum_{j=1}^{n} \left(\lambda_j^2 L_i^f |E_{ji}| e_i(\mathrm{t}) \overline{e_i(\mathrm{t})} + \lambda_i^2 L_j^f |E_{ij}| e_i'(\mathrm{t}) \overline{e_i'(\mathrm{t})} \right)
\end{aligned} \tag{13}$$

Similarly,

$$\begin{aligned}
&\sum_{i=1}^{n} \sum_{j=1}^{n} \lambda_i \mu_i \Re\Big(E_{ij} \left(f_j\left(v_j(\mathrm{t})\right) - f_j\left(\mathrm{w}_j(\mathrm{t})\right)\right) \overline{e_i(\mathrm{t})} \Big) \\
&\leq \frac{1}{2} \sum_{i=1}^{n} \sum_{j=1}^{n} \left(\lambda_j \mu_j L_i^f |E_{ji}| + \lambda_i \mu_i L_j^f |E_{ij}| \right) e_i(\mathrm{t}) \overline{e_i(\mathrm{t})},
\end{aligned} \tag{14}$$

Moreover, computing the derivative of $L_3(\mathrm{t})$, we get

$$
\begin{aligned}
\frac{dL_3(t)}{dt} &= e^{2\epsilon t}\sum_{i=1}^{n}\Bigg[\epsilon\left(\nu_i\theta_i'(t)+\sigma_i\theta_i(t)\right)\overline{\left(\nu_i\theta_i'(t)+\sigma_i\theta_i(t)\right)} \\
&+\frac{1}{2}\left(\nu_i\theta_i''(t)+\sigma_i\theta_i'(t)\right)\overline{\left(\nu_i\theta_i'(t)+\sigma_i\theta_i(t)\right)} \\
&+\frac{1}{2}\left(\nu_i\theta_i'(t)+\sigma_i\theta_i(t)\right)\overline{\left(\nu_i\theta_i''(t)+\sigma_i\theta_i'(t)\right)}\Bigg] \\
&= e^{2\epsilon t}\sum_{i=1}^{n}\Bigg[\left(\epsilon\nu_i^2-\nu_i^2+\nu_i\sigma_i-\nu_i^2\widetilde{\beta}_i\right)\theta_i'(t)\overline{\theta_i'(t)} \\
&+\left(\epsilon\sigma_i^2-\nu_i\sigma_i-\nu_i\sigma_i\widetilde{\alpha}_i\right)\theta_i(t)\overline{\theta_i(t)} \\
&+2\left(\epsilon\nu_i\sigma_i-\frac{1}{2}\nu_i^2-\frac{1}{2}\nu_i\sigma_i+\frac{1}{2}\sigma_i^2-\frac{1}{2}\nu_i^2\widetilde{\alpha}_i-\frac{1}{2}\nu_i\sigma_i\widetilde{\beta}_i\right)\Re(\theta_i'(t)\overline{\theta_i(t)}) \\
&+\nu_i^2\Re\left(\left(f_i(v_i(t))-f_i(w_i(t))\right)\overline{\theta_i'(t)}\right) \\
&+\nu_i\sigma_i\Re\left(\left(f_i(v_i(t))-f_i(w_i(t))\right)\overline{\theta_i(t)}\right).
\end{aligned}
\tag{15}
$$

Similarly to (13)–(15), we have

$$
\begin{aligned}
&\sum_{i=1}^{n}\nu_i^2\Re\left(\left(f_i(v_i(t))-f_i(w_i(t))\right)\overline{\theta_i'(t)}\right) \\
&\leq \frac{1}{2}\sum_{i=1}^{n}\left(\nu_i^2L_i^f e_i(t)\overline{e_i(t)}+\nu_i^2L_i^f\theta_i'(t)\overline{\theta_i'(t)}\right),
\end{aligned}
\tag{16}
$$

and

$$
\begin{aligned}
&\sum_{i=1}^{n}\nu_i\sigma_i\Re\left(\left(f_i(v_i(t))-f_i(w_i(t))\right)\overline{\theta_i(t)}\right) \\
&\leq \frac{1}{2}\sum_{i=1}^{n}\left(\nu_i\sigma_iL_i^f e_i(t)\overline{e_i(t)}+\nu_i\sigma_iL_i^f\theta_i(t)\overline{\theta_i(t)}\right)
\end{aligned}
\tag{17}
$$

From (11)–(17), one has

$$
\begin{aligned}
\frac{dL(\mathrm{t})}{dt} &\le e^{2\epsilon \mathrm{t}}\sum_{i=1}^{n}\Bigg[\left(\epsilon\lambda_i^2-\lambda_i^2 a_i+\lambda_i\mu_i-\lambda_i^2\beta_i+\frac{1}{2}\sum_{j=1}^{n}\lambda_i^2 L_j^f|E_{ij}|\right)e_i'(\mathrm{t})\overline{e_i'(\mathrm{t})}\\
&+\Bigg(\epsilon\mu_i^2+\epsilon\delta_i+\frac{1}{2}\nu_i^2 L_i^f+\frac{1}{2}\nu_i\sigma_i L_i^f\\
&-\lambda_i\mu_i d_i-\lambda_i\mu_i\alpha_i+\frac{1}{2}\sum_{j=1}^{n}\left(\lambda_j^2 L_i^f|E_{ji}|+\lambda_j\mu_j L_i^f|E_{ji}|+\lambda_i\mu_i L_j^f|E_{ij}|\right)\Bigg)e_i(\mathrm{t})\overline{e_i(\mathrm{t})}\\
&+\left(2\epsilon\lambda_i\mu_i+\delta_i-\lambda_i\mu_i a_i+\mu_i^2-\lambda_i^2 d_i-\lambda_i^2\alpha_i-\lambda_i\mu_i\beta_i\right)\Re(e_i'(\mathrm{t})\overline{e_i(\mathrm{t})})\\
&+\left(\epsilon\nu_i^2+\frac{1}{2}\nu_i^2 L_i^f-\nu_i^2+\nu_i\sigma_i-\nu_i^2\widetilde{\beta}_i\right)\theta_i'(\mathrm{t})\overline{\theta_i'(\mathrm{t})}\\
&+\left(\epsilon\sigma_i^2+\epsilon\eta_i+\frac{1}{2}\nu_i\sigma_i L_i^f-\nu_i\sigma_i-\nu_i\sigma_i\widetilde{\alpha}_i\right)\theta_i(\mathrm{t})\overline{\theta_i(\mathrm{t})}\\
&+\left(2\epsilon\nu_i\sigma_i+\eta_i-\nu_i^2-\nu_i\sigma_i+\sigma_i^2-\nu_i^2\widetilde{\alpha}_i-\nu_i\sigma_i\widetilde{\beta}_i\right)\Re(\theta_i'(\mathrm{t})\overline{\theta_i(\mathrm{t})})\Bigg]\\
&= e^{2\epsilon \mathrm{t}}\sum_{i=1}^{n}\Bigg[\Gamma_i e_i'(\mathrm{t})\overline{e_i'(\mathrm{t})}+\Delta_i e_i(\mathrm{t})\overline{e_i(\mathrm{t})}+\frac{\Lambda_i}{2}\left(e_i'(\mathrm{t})\overline{e_i(\mathrm{t})}+\overline{e_i'(\mathrm{t})}e_i(\mathrm{t})\right)\\
&+\widetilde{\Gamma}_i\theta_i'(\mathrm{t})\overline{\theta_i'(\mathrm{t})}+\widetilde{\Delta}_i\theta_i(\mathrm{t})\overline{\theta_i(\mathrm{t})}+\frac{\widetilde{\Lambda}_i}{2}\left(\theta_i'(\mathrm{t})\overline{\theta_i(\mathrm{t})}+\overline{\theta_i'(\mathrm{t})}\theta_i(\mathrm{t})\right)\Bigg]\\
&= e^{2\epsilon \mathrm{t}}\Bigg[\sum_{i=1}^{n}\Gamma_i\left(e_i'(\mathrm{t})+\frac{\Lambda_i}{2\Gamma_i}e_i(\mathrm{t})\right)\overline{\left(e_i'(\mathrm{t})+\frac{\Lambda_i}{2\Gamma_i}e_i(\mathrm{t})\right)}+\sum_{i=1}^{n}\left(\Delta_i-\frac{\Lambda_i^2}{4\Gamma_i}\right)e_i(\mathrm{t})\overline{e_i(\mathrm{t})}\\
&+\sum_{i=1}^{n}\widetilde{\Gamma}_i\left(\theta_i'(\mathrm{t})+\frac{\widetilde{\Lambda}_i}{2\widetilde{\Gamma}_i}\theta_i(\mathrm{t})\right)\overline{\left(\theta_i'(\mathrm{t})+\frac{\widetilde{\Lambda}_i}{2\widetilde{\Gamma}_i}\theta_i(\mathrm{t})\right)}+\sum_{i=1}^{n}\left(\widetilde{\Delta}_i-\frac{\widetilde{\Lambda}_i^2}{4\widetilde{\Gamma}_i}\right)\theta_i(\mathrm{t})\overline{\theta_i(\mathrm{t})}\Bigg]\\
&\le 0.
\end{aligned}
\tag{18}
$$

Therefore, $L(\mathrm{t}) \le L(t_0)$, for all $t \ge t_0$ and

$$\|e(\mathrm{t})\|^2+\|\theta(\mathrm{t})\|^2 \le \frac{2L(t_0)}{\min\{\delta_{min},\eta_{min}\}}e^{-2\epsilon \mathrm{t}},$$

where $\delta_{min} = \min\limits_{1\le i\le n,}\{\delta_i\}$ and $\eta_{min} = \min\limits_{1\le i\le n,}\{\eta_i\}$.
Therefore,

$$\frac{1}{2}\left(\|e(\mathrm{t})\|+\|\theta(\mathrm{t})\|\right)^2 \le \frac{2L(t_0)}{\min\{\delta_{min},\eta_{min}\}}e^{-2\epsilon \mathrm{t}},$$

we can obtain

$$\|e(\mathrm{t})\|+\|\theta(\mathrm{t})\| \le \sqrt{\frac{4L(t_0)}{\min\{\delta_{min},\eta_{min}\}}}e^{-\epsilon \mathrm{t}}.$$

Based on Definition 1, response system (5) and drive system (3) can achieve global ExpS at the rate ϵ under controller (8). The proof is completed.

Evidently, when $\delta_i = -\lambda_i \mu_i (2\epsilon - a_i - \beta_i) - \mu_i^2 + \lambda_i^2 (d_i + \alpha_i) > 0$, then $\Lambda_i = 0$ and, if $\eta_i = -2\epsilon \nu_i \sigma_i + \nu_i^2 + \nu_i \sigma_i - \sigma_i^2 + \nu_i^2 \widetilde{\alpha}_i + \nu_i \sigma_i \widetilde{\beta}_i > 0$ therefore $\widetilde{\Lambda}_i = 0$. Given this assumption, condition **(H.2)** can be restated as follows:

(H.3) For every $i = 1, \cdots, n$, there exist nonzero constants λ_i, μ_i, ν_i, σ_i and positive number ϵ such that

$$\Gamma_i < 0, \quad \Delta_i < 0,$$

and

$$\widetilde{\Gamma}_i < 0, \quad \widetilde{\Delta}_i < 0.$$

Theorem 2 *By applying the feedback controller (8) and under* **(H.1)** *and* **(H.3)**, *the neural systems (3) and (5) achieve ExpS.*

Remark 1 In **(H.2)**, numerous parameters are introduced in order to improve the quality of the results obtained. To simplify the application of these conditions, it is possible to select appropriate parameters. Thus, Theorem 2 is proposed, offering criteria that are easily verifiable in practice.

4 Numerical Simulations

In this section, we illustrate the effectiveness of the proposed method using a numerical example. Consider the following driving InCVCoNNs containing two complex-valued neurons

$$\begin{cases} \text{STM}: \mathrm{w}_i''(\mathrm{t}) + a_i \mathrm{w}_i'(\mathrm{t}) = -d_i \mathrm{w}_i(\mathrm{t}) + \sum_{j=1}^{2} E_{ij} f_j\left(\mathrm{w}_j(\mathrm{t})\right) + \sum_{j=1}^{2} H_{ij} f_j\left(\mathrm{w}_j(\kappa \mathrm{t})\right) \\ \qquad\qquad + c_i s_i(\mathrm{t}) + I_i, \\ \text{LTM}: s_i''(\mathrm{t}) + s_i'(\mathrm{t}) \quad = -s_i(\mathrm{t}) + f_i(\mathrm{w}_i(\mathrm{t})), \end{cases} \tag{19}$$

and the controlled response inertial InCVCoNNs is represented as follows

$$\begin{cases} \text{STM}: v_i''(\mathrm{t}) + a_i v_i'(\mathrm{t}) = -d_i v_i(\mathrm{t}) + \sum_{j=1}^{2} E_{ij} f_j\left(v_j(\mathrm{t})\right) + \sum_{j=1}^{2} H_{ij} f_j\left(v_j(\kappa \mathrm{t})\right) \\ \qquad\qquad + c_i q_i(\mathrm{t}) + I_i + \mathcal{U}_i(\mathrm{t}), \\ \text{LTM}: q_i''(\mathrm{t}) + q_i'(\mathrm{t}) \quad = -q_i(\mathrm{t}) + f_i(v_i(\mathrm{t})) + \mathcal{V}_i(\mathrm{t}), \end{cases} \tag{20}$$

where $i = 1, 2$, $a_1 = 2$, $a_2 = 3$, $d_1 = 4$, $d_2 = 3$, $c_1 = c_2 = 2.5$, $I_1 = 2$, $I_2 = -1$, $\kappa = 0.5$ $f_j(\mathrm{w}_j) = \tanh(\Re(\mathrm{w}_j)) + i\sin(\Im(\mathrm{w}_j))$, $E_{11} = 1 - i$, $E_{12} = -0.3 - 0.5i$, $E_{21} = -1 - 2i$, $E_{22} = 1.5 - i$, $H_{11} = -1.5 + 1.2i$, $H_{12} = -0.2 - 0.3i$, $H_{21} = -0.1 + 0.5i$, $H_{22} = -0.5 - 0.2i$.

On the other hand, it is easy to verify that $L_1^f = L_2^f = 2$. Choose $\epsilon = 0.1$, $\lambda_1 = \lambda_2 = \mu_1 = \mu_2 = 1, \alpha_1 = \alpha_2 = 8, \widetilde{\alpha}_1 = \widetilde{\alpha}_2 = 10, \beta_1 = \beta_2 = 3, \widetilde{\beta}_1 = \widetilde{\beta}_2 = 15, \nu_1 = \sigma_1 = 0.1, \delta_1 = \delta_2 = 16, \eta_1 = \eta_2 = 0.2$. After a few simple manipulations, we obtain:

$$\begin{cases} \Gamma_1 &= -1.9027 < 0, \\ \Gamma_2 &= -0.8612 < 0, \\ \Delta_1 &= -0.9821, \\ \Delta_2 &= -0.4693, \\ \Lambda_1 &= 0.2000, \\ \Lambda_2 &= 0.2000. \end{cases}$$

Therefore, $\Lambda_1^2 = 0.0400 < 4\Gamma_1\Delta_1 = 7.4746$ and $\Lambda_2^2 = 0.0400 < 4\Gamma_2\Delta_2 = 1.6166$. In addition,

$$\begin{cases} \widetilde{\Gamma}_1 &= -0.1390 < 0, \\ \widetilde{\Gamma}_2 &= -0.1390 < 0, \\ \widetilde{\Delta}_1 &= -0.0790, \\ \widetilde{\Delta}_2 &= -0.0790, \\ \widetilde{\Lambda}_1 &= 0.0330, \\ \widetilde{\Lambda}_2 &= 0.0330. \end{cases}$$

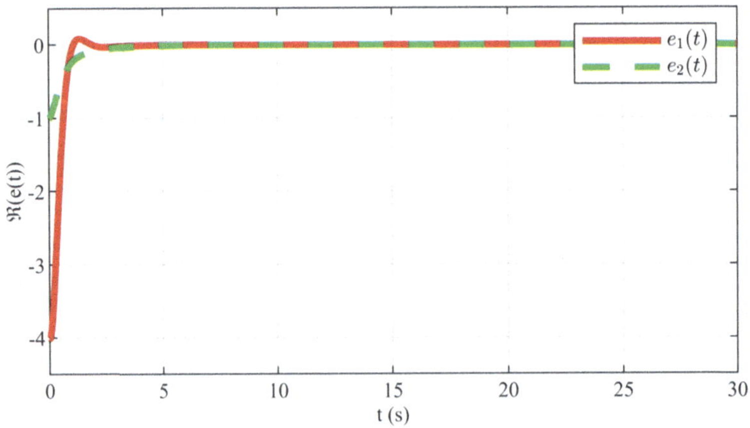

Fig. 1 Time evolution of synchronization errors $\Re(e(\mathrm{t}))$

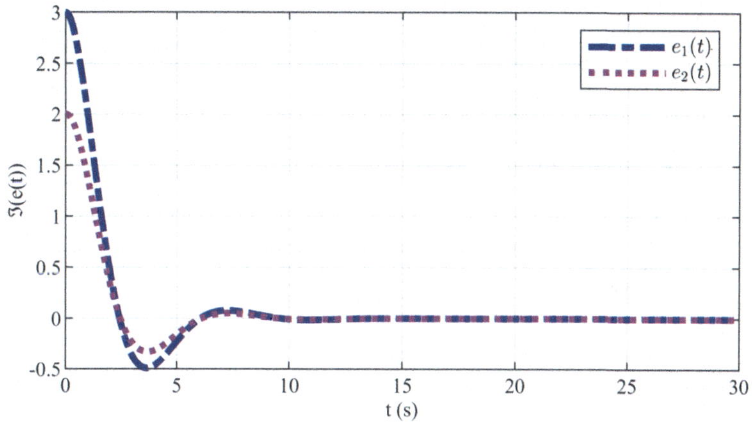

Fig. 2 Time evolution of synchronization errors $\Im(e(t))$

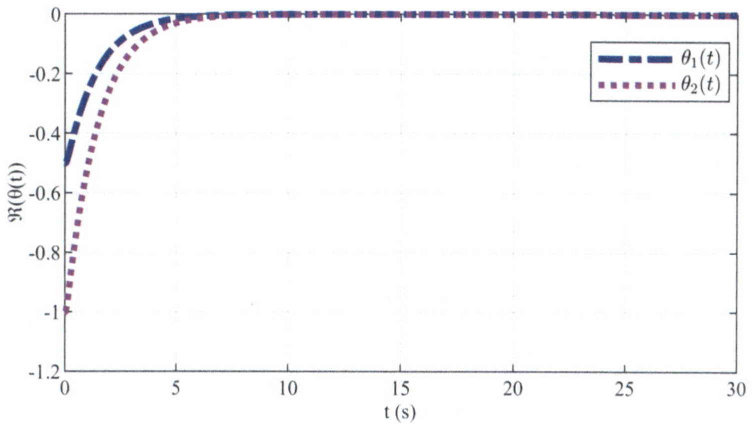

Fig. 3 Time evolution of synchronization errors $\Re(\theta(\mathrm{t}))$

Then, $\widetilde{\Lambda}_1^2 = 0.0011 < 4\widetilde{\Gamma}_1\widetilde{\Delta}_1 = 0.5464$ and $\widetilde{\Lambda}_2^2 = 0.0011 < 4\widetilde{\Gamma}_2\widetilde{\Delta}_2 = 0.2608$. As shown in Figs. 1, 2, 3 and 4, the response system (19) and the drive system (20) achieve ExpS under the feedback controller (8), since all the conditions of Theorem 1 are satisfied.

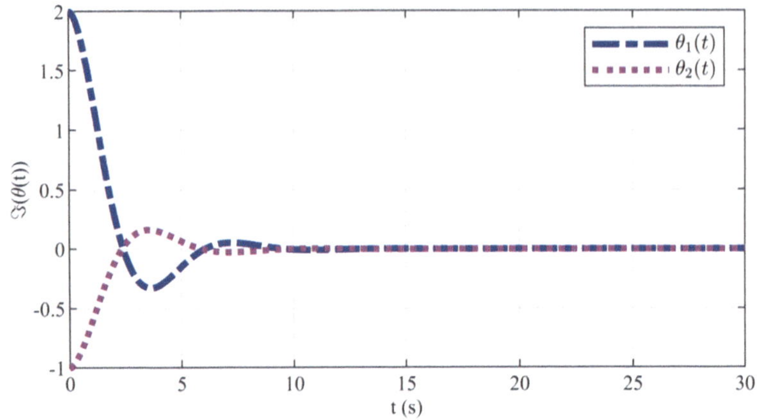

Fig. 4 Time evolution of synchronization errors $\Im(\theta(\mathrm{t}))$

5 Conclusion

This chapter investigates the problem of ExpS in a class of delayed inertial complex-valued InCVCoNNs by employing a direct analytical framework combined with a non-separation strategy. The main contributions of this study can be summarized as follows. First, a novel formulation of generalized InCVCoNNs with proportional delays is introduced. Second, the proposed analysis departs from the conventional reduced-order techniques commonly used for ExpS evaluation. Finally, a direct complex-domain control methodology is developed, allowing the stability analysis to be conducted without relying on classical separation procedures.

Acknowledgements This work was supported by Hunan Provincial Natural Science Foundation Department Joint Fund (grant number 2025JJ80254).

Competing Interests The authors have no conflicts of interest to declare that are relevant to the content of this chapter.

References

1. Kumar, P., Lee, T.H., Erturk, V.S.: A novel two-delayed tri-neuron neural network with an incomplete connection. Nonlinear Dyn. **112**(22), 20269–20293 (2024)
2. Alimi, A.M., Aouiti, C., Assali, E.A.: Finite-time and fixed-time synchronization of a class of inertial neural networks with multi-proportional delays and its application to secure communication. Neurocomputing **332**, 29–43 (2019)
3. Wen, S., Zeng, Z., Huang, T., Meng, Q., Yao, W.: Lag synchronization of switched neural networks via neural activation function and applications in image encryption. IEEE Trans. Neural Netw. Learn. Syst. **26**(7), 1493–1502 (2015)

4. Wan, Y., Zhou, L., Han, J.: Global polynomial synchronization of proportional delay memristive neural networks with uncertain parameters and its application to image encryption. Eng. Appl. Artif. Intell. **147**, 110290 (2025)
5. Kumar, P., Lee, T.H., Erturk, V.S.: A fractional-order multi-delayed bicyclic crossed neural network: Stability, bifurcation, and numerical solution. Neural Netw. **188**, 107436 (2025)
6. Kumar, P., Lee, T.H., Erturk, V.S.: A novel fractional-order neutral-type two-delayed neural network: Stability, bifurcation, and numerical solution. Math. Comput. Simul. **232**, 245–260 (2025)
7. Sivalingam, S.M., Kumar, P., Govindaraj, V.: A Chebyshev neural network-based numerical scheme to solve distributed-order fractional differential equations. Comput. Math. Appl. **164**, 150–165 (2024)
8. Liu, Y., Yang, Y., Liang, T., Li, L.: Existence and global exponential stability of anti-periodic solutions for competitive neural networks with delays in the leakage terms on time scales. Neurocomputing **133**, 471–482 (2014)
9. Zhou, L., Zhao, Z.: Exponential stability of a class of competitive neural networks with multi-proportional delays. Neural Process. Lett. **44**(3), 651–663 (2016)
10. Assali, E.A., Kumar, P., Lee, T.H.: Novel Predefined-time Stability Theorems With Applications to the Synchronization of Competitive Neural Networks With Time-varying Delays. Int. J. Control Autom. Syst. **23**(10), 2821–2836 (2025)
11. Shi, M., Guo, J., Fang, X., Huang, C.: Global exponential stability of delayed inertial competitive neural networks. Adv. Differ. Equ. **2020**(1), 87 (2020)
12. Meyer-Bäse, A., Ohl, F., Scheich, H.: Singular perturbation analysis of competitive neural networks with different time scales. Neural Comput. **8**(8), 1731–1742 (1996)
13. Bondarenko, V.E.: Information processing, memories, and synchronization in chaotic neural network with time delay. Complexity **11**(2), 39–52 (2005)
14. Yang, W., Wang, Y.W., Morărescu, I.C., Liu, X.K., Huang, Y.: Fixed-time synchronization of competitive neural networks with multiple time scales. IEEE Trans. Neural Netw. Learn. Syst. **33**(8), 4133–4138 (2021)
15. Yang, W., Wang, Y.W., Shen, Y., Pan, L.: Cluster synchronization of coupled delayed competitive neural networks with two time scales. Nonlinear Dyn. **90**(4), 2767–2782 (2017)
16. Gan, Q.: Synchronization of competitive neural networks with different time scales and time-varying delay based on delay partitioning approach. Int. J. Mach. Learn. Cybern. **4**(4), 327–337 (2013)
17. Yang, X., Cao, J., Long, Y., Rui, W.: Adaptive lag synchronization for competitive neural networks with mixed delays and uncertain hybrid perturbations. IEEE Trans. Neural Netw. **21**(10), 1656–1667 (2010)
18. Kumar, P., Assali, E.A.: Fixed-time synchronization of fractional-order Hopfield neural networks with proportional delays. Math. Comput. Simul. (2025). https://doi.org/10.1016/j.matcom.2025.07.035
19. Babcock, K.L., Westervelt, R.M.: Stability and dynamics of simple electronic neural networks with added inertia. Physica D **23**(1–3), 464–469 (1986)
20. Li, Y., Qin, J.: Existence and global exponential stability of anti-periodic solutions for generalised inertial competitive neural networks with time-varying delays. J. Exp. Theor. Artif. Intell. **32**(2), 291–307 (2020)
21. Ke, Y., Miao, C.: Stability analysis of inertial Cohen-Grossberg-type neural networks with time delays. Neurocomputing **117**, 196–205 (2013)
22. Ke, Y., Miao, C.: Stability and existence of periodic solutions in inertial BAM neural networks with time delay. Neural Comput. Appl. **23**(3), 1089–1099 (2013)
23. Ke, Y., Miao, C.: Anti-periodic solutions of inertial neural networks with time delays. Neural Process. Lett. **45**(2), 523–538 (2017)
24. Hu, J., Cao, J., Alofi, A., Al-Mazrooei, A., Elaiw, A.: Pinning synchronization of coupled inertial delayed neural networks. Cogn. Neurodyn. **9**(3), 341–350 (2015)
25. Li, X., Li, X., Hu, C.: Some new results on stability and synchronization for delayed inertial neural networks based on non-reduced order method. Neural Netw. **96**, 91–100 (2017)

26. Aouiti, C., Assali, E.A.: Nonlinear Lipschitz measure and adaptive control for stability and synchronization in delayed inertial Cohen-Grossberg-type neural networks. Int. J. Adapt. Control Signal Process. **33**(10), 1457–1477 (2019)
27. Shi, M., Guo, J., Fang, X., Huang, C.: Global exponential stability of delayed inertial competitive neural networks. Adv. Differ. Equ. **2020**(1), 87 (2020)
28. Huang, C., Liu, B.: New studies on dynamic analysis of inertial neural networks involving non-reduced order method. Neurocomputing **325**, 283–287 (2019)
29. Li, Y., Qin, J.: Existence and global exponential stability of anti-periodic solutions for generalised inertial competitive neural networks with time-varying delays. J. Exp. Theor. Artif. Intell. **32**(2), 291–307 (2020)
30. Aizenberg, I.: Multiple-valued logic and complex-valued neural networks. In: Moraga, C. (ed.) A Passion for Multi-Valued Logic and Soft Computing, 153–171. Springer, Cham (2016)
31. Tanaka, G., Aihara, K.: Complex-valued multistate associative memory with nonlinear multilevel functions for gray-level image reconstruction. IEEE Trans. Neural Netw. **20**(9), 1463–1473 (2009)
32. Yu, J., Hu, C., Jiang, H., Wang, L.: Exponential and adaptive synchronization of inertial complex-valued neural networks: A non-reduced order and non-separation approach. Neural Netw. **124**, 50–59 (2020)
33. Xu, C., Jiang, M., Hu, J.: Fixed-time synchronization of complex-valued memristive competitive neural networks based on two novel fixed-time stability theorems. Neural Comput. Appl. **35**(30), 22605–22620 (2023)
34. Xu, D., Tan, M.: Multistability of delayed complex-valued competitive neural networks with discontinuous non-monotonic piecewise nonlinear activation functions. Commun. Nonlinear Sci. Numer. Simul. **62**, 352–377 (2018)
35. Zhao, Y., Ren, S.: Synchronization for a class of complex-valued memristor-based competitive neural networks with different time scales. Neural Netw. **24**(6), 15 (2021)
36. Yu, J., Hu, C., Jiang, H., Wang, L.: Exponential and adaptive synchronization of inertial complex-valued neural networks: A non-reduced order and non-separation approach. Neural Netw. **124**, 50–59 (2020)
37. Wu, A., Zeng, Z.: Exponential passivity of memristive neural networks with time delays. Neural Netw. **49**, 11–18 (2014)
38. Akca, H., Aouiti, C., Touati, F., Xu, C.: Finite-time passivity of neutral-type complex-valued neural networks with time-varying delays. Math. Biosci. Eng. **21**(5), 6097–6122 (2024)
39. Shi, X., Wang, Z., Han, L.: Finite-time stochastic synchronization of time-delay neural networks with noise disturbance. Nonlinear Dyn. **88**(4), 2747–2755 (2017)
40. Zhou, L.: On the global dissipativity of a class of cellular neural networks with multipantograph delays. Adv. Artif. Neural Syst. **2011**(1), 941426 (2011)

Mathematical Foundations and Optimization Techniques in Deep Neural Networks

Dhilshath Shajahan

Abstract Deep learning has become a revolutionary force in the field of artificial intelligence, demonstrating exceptional performance across multiple application areas such as image analysis, text understanding, and audio processing. However, the theoretical understanding of why these models work so effectively remains an active area of research. This chapter provides a comprehensive exploration of the mathematical foundations underlying neural network optimization, examining both classical and contemporary techniques that drive efficient training of deep learning architectures. We present rigorous mathematical formulations, analyze convergence properties, and discuss practical implications of various optimization strategies. The chapter emphasizes the critical role of mathematical principles in addressing challenges such as vanishing gradients, slow convergence, and generalization in high-dimensional parameter spaces.

Keywords Neural networks · Mathematical optimization · Gradient descent · Adaptive learning rates · Convergence analysis · Deep learning

1 Introduction

The rapid advancement of deep learning has revolutionized numerous fields, from autonomous vehicles to medical diagnosis and natural language understanding [14]. At the core of these successes lies the ability to effectively train neural networks with millions or even billions of parameters. The training process fundamentally relies on optimization techniques that iteratively adjust network parameters to minimize a loss function that quantifies the discrepancy between predicted and actual outputs [8]. Despite the empirical success of deep learning, many aspects remain poorly understood from a theoretical perspective [26]. Questions regarding why certain architectures generalize well, how optimization algorithms navigate complex loss

D. Shajahan (✉)
Department of Mathematics, Sri Sairam Engineering College, Sai Leo Nagar, West Tambaram,Chennai 600044, India
e-mail: dhilshajahan@gmail.com

L. Kaur and P. Kumar (eds.), *Neural Networks: Theory, Algorithms, Simulation, and Applications*, Studies in Computational Intelligence 1273,
https://doi.org/10.1007/978-3-032-18750-5_4

landscapes, and what mathematical properties guarantee convergence are subjects of ongoing investigation. Understanding these mathematical foundations is essential not only for advancing theoretical knowledge but also for developing more efficient and reliable deep learning systems [2].

The foundations of neural network optimization trace back to the seminal work on backpropagation by Rumelhart et al. [22], which enabled efficient gradient computation through multi-layer networks. This breakthrough, combined with advances in computational power and data availability, has led to the deep learning revolution we witness today. Early optimization methods relied primarily on gradient descent and its stochastic variants [3], which remain fundamental to modern practice. The development of convolutional neural networks by LeCun et al. [15] demonstrated the power of specialized architectures for image recognition tasks, while also highlighting the importance of effective optimization strategies. More recently, innovations in network architecture, such as residual networks [9], have been driven by insights into optimization dynamics and gradient flow through deep networks.

Traditional optimization theory, developed primarily for convex problems [20], often fails to fully explain the behavior of neural network training, which involves non-convex optimization in extremely high-dimensional spaces. Modern neural networks typically operate in parameter spaces with dimensionality ranging from thousands to billions, presenting unique computational and theoretical challenges [4]. The non-convex nature of neural network loss functions introduces numerous local minima, saddle points, and flat regions that complicate the optimization process [16]. Recent theoretical work has shown that, despite non-convexity, gradient descent can efficiently escape saddle points and converge to high-quality solutions [5]. Furthermore, the relationship between training performance and generalization to unseen data remains incompletely characterized by classical statistical learning theory [1].

The field has witnessed remarkable progress in optimization algorithms specifically designed for deep learning. The introduction of adaptive learning rate methods, particularly Adam [13], has significantly improved training efficiency and stability across diverse architectures and tasks. These methods automatically adjust per-parameter learning rates based on gradient history, reducing the need for extensive hyperparameter tuning. Momentum-based approaches [24] have proven effective in accelerating convergence by accumulating velocity in consistent gradient directions while dampening oscillations. The interplay between optimization algorithms and architectural innovations, such as batch normalization [10], has been crucial in enabling training of increasingly deep networks. Recent research has also highlighted the importance of implicit biases in optimization algorithms. For instance, stochastic gradient descent tends to find flat minima that generalize better than sharp minima [12], suggesting that the optimization algorithm itself acts as an implicit regularizer beyond explicit regularization techniques [23].

Several theoretical frameworks have emerged to understand deep learning optimization. The Neural Tangent Kernel theory [11] provides convergence guarantees for infinitely wide networks, offering insights into the training dynamics of overparameterized models. Work on the double descent phenomenon [1] has challenged traditional understanding of the bias-variance tradeoff, showing that test error can decrease again after the interpolation threshold. Classical learning theory, as presented in foundational texts [2], provides statistical bounds on generalization error based on model complexity. However, modern deep networks often defy these classical bounds, achieving excellent generalization despite having far more parameters than training samples [26]. This has motivated research into alternative explanations based on implicit regularization, optimization dynamics, and the geometry of loss landscapes [16].

Beyond theoretical understanding, practical optimization in deep learning requires careful consideration of numerous factors. Learning rate scheduling strategies have proven essential for achieving optimal performance, with techniques ranging from simple step decay to sophisticated warm restart schemes becoming standard practice. Initialization strategies [7] play a crucial role in ensuring stable gradient flow during early training phases. Regularization techniques, including weight decay and dropout [23], remain important tools for controlling overfitting and improving generalization. The interaction between these regularization methods and optimization algorithms, particularly regarding weight decay implementation in adaptive methods [17], has been the subject of recent investigation. For large-scale applications, distributed optimization strategies enable training on datasets and models that would be intractable on single machines. Data parallelism and model parallelism approaches, combined with efficient communication protocols, have facilitated the training of models with hundreds of billions of parameters.

Despite significant progress, several fundamental questions remain open. Understanding the precise conditions under which gradient-based methods converge to global or near-global minima in non-convex settings continues to be an active research area [5]. The development of tighter, non-vacuous generalization bounds that explain deep learning success represents another major challenge [26]. The computational cost of training large models has motivated research into more efficient optimization algorithms, including second-order methods that could potentially achieve faster convergence [20]. However, scaling these methods to modern problem sizes while maintaining computational feasibility remains challenging. Meta-learning approaches that learn optimization algorithms themselves, rather than hand-crafting them, represent a promising direction for future research. Similarly, neural architecture search, which automates the design of network structures, relies heavily on efficient optimization techniques.

This chapter addresses these challenges by presenting a systematic treatment of optimization in neural networks, bridging classical optimization theory with modern deep learning practice. We examine how mathematical principles guide the design of effective optimization algorithms and how these algorithms can be analyzed theoretically. The remainder of this chapter is organized as follows. Section 2 establishes the mathematical framework for neural network optimization, including problem formulation and loss functions. Section 3 examines classical gradient-based optimization methods and their variants, including stochastic gradient descent [3] and momentum-based approaches [24]. Section 4 explores advanced adaptive optimization techniques such as Adam [13] and second-order methods [20]. Section 5 discusses convergence analysis and generalization properties, drawing on recent theoretical advances [5].

Section 6 presents regularization techniques from a mathematical perspective, including dropout [23] and batch normalization [10]. Section 7 addresses practical considerations in implementation. Finally, Sect. 10 concludes with future research directions and open problems in the field.

Throughout this chapter, we emphasize the interplay between theoretical understanding and practical performance, providing both rigorous mathematical formulations and insights into why certain techniques work well in practice. Our goal is to equip readers with both the theoretical foundations and practical knowledge necessary to effectively apply and advance optimization techniques in deep learning.

2 Mathematical Framework for Neural Network Optimization

2.1 *Problem Formulation*

Consider a supervised learning task where we have a dataset $\mathcal{D} = \{(x_i, y_i)\}_{i=1}^N$ with input-output pairs, where $x_i \in \mathbb{R}^d$ represents the input features and $y_i \in \mathbb{R}^m$ (or $y_i \in \{1, \ldots, C\}$ for classification) represents the target output. A neural network defines a parametric function $f_\theta : \mathbb{R}^d \to \mathbb{R}^m$, where $\theta \in \mathbb{R}^p$ represents the collection of all network parameters including weights and biases [2].

The optimization problem in neural network training can be formally stated as:

$$\theta^* = \arg\min_{\theta \in \mathbb{R}^p} \ \mathcal{L}(\theta) = \frac{1}{N} \sum_{i=1}^{N} \ell(f_\theta(x_i), y_i) \tag{1}$$

where $\ell : \mathbb{R}^m \times \mathbb{R}^m \to \mathbb{R}$ is a loss function measuring the discrepancy between predictions and true labels, and $\mathcal{L}(\theta)$ represents the empirical risk.

2.2 *Loss Functions*

The selection of an appropriate loss function varies based on the problem type. In regression tasks, Mean Squared Error (MSE) is frequently used:

$$\mathcal{L}_{\text{MSE}}(\theta) = \frac{1}{N} \sum_{i=1}^{N} \|y_i - f_\theta(x_i)\|^2 \tag{2}$$

The MSE loss is differentiable and convex with respect to the network output, making it amenable to gradient-based optimization. However, it can be sensitive to outliers due to the quadratic penalty.

For classification tasks, the cross-entropy loss is preferred:

$$\mathcal{L}_{\mathrm{CE}}(\theta) = -\frac{1}{N}\sum_{i=1}^{N}\sum_{c=1}^{C} y_{i,c}\log(f_\theta(x_i)_c) \tag{3}$$

where $y_{i,c}$ is a binary indicator for class membership and $f_\theta(x_i)_c$ represents the predicted probability for class c. The cross-entropy loss directly optimizes the log-likelihood of the correct class, providing stronger gradients for misclassified examples compared to MSE.

2.3 *The Loss Landscape*

Understanding the geometry of the loss landscape $\mathcal{L}(\theta)$ is crucial for analyzing optimization algorithms [16]. For neural networks, this landscape is highly non-convex, characterized by:

1. **Multiple local minima**: Unlike convex optimization, the loss function may have numerous local minima with varying quality.
2. **Saddle points**: Points where the gradient vanishes but the Hessian has both positive and negative eigenvalues, potentially slowing optimization [21].
3. **Flat regions**: Areas where gradients are very small, making progress difficult.
4. **Sharp minima versus flat minima**: Recent work suggests that flat minima, where the loss changes slowly around the optimum, tend to generalize better than sharp minima [12].

2.4 *Neural Network Architectures*

A multilayer feedforward neural network comprising L layers can be mathematically expressed as a nested composition of transformations:

$$f_\theta(x) = h^{(L)} \circ h^{(L-1)} \circ \cdots \circ h^{(1)}(x) \tag{4}$$

where each individual layer $h^{(l)}$ implements a linear transformation followed by element-wise application of a nonlinear activation:

$$h^{(l)}(z) = \sigma(W^{(l)}z + b^{(l)}) \tag{5}$$

Here, $W^{(l)} \in \mathbb{R}^{n_l \times n_{l-1}}$ denotes the learnable weight matrix connecting layers $l-1$ and l, $b^{(l)} \in \mathbb{R}^{n_l}$ represents the bias term, and $\sigma(\cdot)$ designates a nonlinear activation function that introduces crucial nonlinearity into the model.

2.4.1 Activation Functions

Several activation functions have emerged as standard choices in deep learning architectures, each with distinct mathematical properties.

Rectified Linear Unit (ReLU): This piecewise linear function is defined as:

$$\sigma(z) = \max(0, z) \tag{6}$$

The corresponding derivative is given by $\sigma'(z) = \mathbb{1}_{z>0}$, which equals unity for positive inputs and zero otherwise.

Sigmoid Function: The logistic sigmoid provides a smooth, bounded activation:

$$\sigma(z) = \frac{1}{1 + e^{-z}} \tag{7}$$

Its derivative exhibits the convenient form $\sigma'(z) = \sigma(z)(1 - \sigma(z))$, facilitating efficient backpropagation.

Hyperbolic Tangent (Tanh): This function maps inputs to the range $(-1, 1)$:

$$\sigma(z) = \tanh(z) = \frac{e^z - e^{-z}}{e^z + e^{-z}} \tag{8}$$

The derivative is expressed as $\sigma'(z) = 1 - \tanh^2(z)$, demonstrating symmetry around the origin.

Among these options, ReLU has gained widespread adoption in modern architectures due to two principal advantages: its straightforward computational implementation requiring only a thresholding operation, and its effectiveness in alleviating gradient attenuation issues that plague deeper networks [7]. Unlike sigmoid and tanh activations, which can suffer from saturation in their flat regions, ReLU maintains a constant gradient for positive inputs, thereby facilitating more effective gradient flow during backpropagation through deep architectures.

3 Classical Gradient-Based Optimization

3.1 Gradient Descent

The basic method for minimizing $\mathcal{L}(\theta)$ is gradient descent (GD), which repeatedly adjusts parameters along the direction of steepest descent [22]:

$$\theta_{t+1} = \theta_t - \eta \nabla_\theta \mathcal{L}(\theta_t) \tag{9}$$

where $\eta > 0$ is the learning rate (or step size), and $\nabla_\theta \mathcal{L}(\theta_t)$ denotes the gradient of the loss function with respect to parameters at iteration t.

The gradient is computed using the backpropagation algorithm, which efficiently calculates gradients by applying the chain rule recursively through the network layers:

$$\nabla_\theta \mathcal{L}(\theta) = \frac{1}{N} \sum_{i=1}^{N} \nabla_\theta \ell(f_\theta(x_i), y_i) \tag{10}$$

3.1.1 Convergence Analysis

For convex functions with Lipschitz-continuous gradients, gradient descent with appropriate step size $\eta \leq 1/L$ (where L is the Lipschitz constant) guarantees convergence to the global minimum. Specifically, after T iterations:

$$\mathcal{L}(\theta_T) - \mathcal{L}(\theta^*) \leq \frac{\|\theta_0 - \theta^*\|^2}{2\eta T} \tag{11}$$

However, neural network loss functions are non-convex, making such guarantees inapplicable. Nevertheless, gradient descent often finds good local minima in practice.

3.2 Stochastic Gradient Descent

Computing the full gradient in Eq. (10) requires evaluating the loss over the entire dataset, which is computationally prohibitive for large-scale problems. Stochastic Gradient Descent (SGD) addresses this by approximating the gradient using a single randomly selected sample [3]:

$$\theta_{t+1} = \theta_t - \eta \nabla_\theta \ell(f_\theta(x_i), y_i) \tag{12}$$

where (x_i, y_i) is randomly sampled from $\mathcal{D}$ at each iteration.

The stochastic gradient is an unbiased estimator of the true gradient:

$$\mathbb{E}[\nabla_\theta \ell(f_\theta(x_i), y_i)] = \nabla_\theta \mathcal{L}(\theta) \tag{13}$$

While SGD introduces noise into the optimization process, this noise can actually be beneficial by helping escape sharp local minima and saddle points.

3.3 Mini-batch Gradient Descent

A practical compromise between full-batch gradient descent and SGD is mini-batch gradient descent, which computes gradients over small batches of samples [4]:

$$\theta_{t+1} = \theta_t - \eta \frac{1}{B} \sum_{i \in \mathcal{B}_t} \nabla_\theta \ell(f_\theta(x_i), y_i) \tag{14}$$

where $\mathcal{B}_t$ is a mini-batch of size B randomly sampled at iteration t.

Mini-batch gradient descent offers several advantages:

- Reduced variance compared to pure SGD
- Computational efficiency through parallelization
- Better utilization of modern hardware architectures (GPUs)
- Smoother convergence trajectories.

Typical batch sizes range from 32 to 512, balancing computational efficiency with gradient estimate quality.

3.4 Momentum-Based Methods

Standard gradient descent can oscillate in directions with high curvature while making slow progress along directions with low curvature. Momentum methods address this by accumulating a velocity vector that dampens oscillations and accelerates convergence [24].

3.4.1 Classical Momentum

The momentum method introduces an exponentially weighted moving average of past gradients:

$$v_{t+1} = \beta v_t + (1 - \beta) \nabla_\theta \mathcal{L}(\theta_t) \tag{15}$$

$$\theta_{t+1} = \theta_t - \eta v_{t+1} \tag{16}$$

where $\beta \in [0, 1)$ is the momentum coefficient, typically set to 0.9 or 0.99.

The momentum term v_t can be interpreted as velocity in parameter space. When gradients consistently point in the same direction, the velocity accumulates, accelerating movement. When gradients oscillate, the velocity dampens these oscillations.

3.4.2 Nesterov Accelerated Gradient

Nesterov momentum provides a more sophisticated update by looking ahead to where the momentum term will take us [19]:

$$v_{t+1} = \beta v_t + \nabla_\theta \mathcal{L}(\theta_t - \eta \beta v_t) \tag{17}$$

$$\theta_{t+1} = \theta_t - \eta v_{t+1} \tag{18}$$

This "look-ahead" mechanism allows the algorithm to correct its course before making large steps in the wrong direction, often resulting in faster convergence than classical momentum.

4 Advanced Adaptive Optimization Techniques

4.1 Adaptive Learning Rate Methods

A major limitation of standard gradient descent methods is their use of a single global learning rate for all parameters. Adaptive methods address this by maintaining per-parameter learning rates that adapt based on the history of gradients.

4.2 AdaGrad

AdaGrad (Adaptive Gradient Algorithm) adjusts the learning rate for each parameter separately through inverse scaling with respect to the accumulated squared gradient history:

$$G_t = G_{t-1} + (\nabla_\theta \mathcal{L}(\theta_t))^2 \tag{19}$$

$$\theta_{t+1} = \theta_t - \frac{\eta}{\sqrt{G_t + \epsilon}} \odot \nabla_\theta \mathcal{L}(\theta_t) \tag{20}$$

In this formulation, $\odot$ denotes element-wise multiplication, $G_t \in \mathbb{R}^p$ accumulates the sum of squared gradients across iterations, and ϵ (typically set to 10^{-8}) serves as a numerical stability constant to avoid division by zero. AdaGrad automatically decreases the learning rate for parameters with large gradients and increases it for parameters with small gradients. This is particularly beneficial for sparse features. However, the continual accumulation of squared gradients causes the learning rate to decrease monotonically, potentially stopping learning prematurely.

4.3 RMSProp

RMSProp mitigates AdaGrad's problem of excessively decreasing learning rates through the use of an exponential moving average of squared gradients:

$$v_t = \gamma v_{t-1} + (1 - \gamma)(\nabla_\theta \mathcal{L}(\theta_t))^2 \tag{21}$$

$$\theta_{t+1} = \theta_t - \frac{\eta}{\sqrt{v_t + \epsilon}} \odot \nabla_\theta \mathcal{L}(\theta_t) \tag{22}$$

where $\gamma \in [0, 1)$ is the decay rate, typically set to 0.9.

By using an exponentially weighted moving average, RMSProp maintains a running estimate of gradient magnitudes that adapts to recent history, allowing the learning rate to increase again when appropriate.

4.4 Adam

The Adam optimizer, which stands for Adaptive Moment Estimation, integrates concepts from both momentum methods and RMSProp to achieve efficient optimization [13]. This algorithm maintains running averages of both the gradient itself and the squared gradient:

$$m_t = \beta_1 m_{t-1} + (1 - \beta_1)\nabla_\theta \mathcal{L}(\theta_t) \tag{23}$$

$$v_t = \beta_2 v_{t-1} + (1 - \beta_2)(\nabla_\theta \mathcal{L}(\theta_t))^2 \tag{24}$$

$$\hat{m}_t = \frac{m_t}{1 - \beta_1^t} \tag{25}$$

$$\hat{v}_t = \frac{v_t}{1 - \beta_2^t} \tag{26}$$

$$\theta_{t+1} = \theta_t - \frac{\eta}{\sqrt{\hat{v}_t} + \epsilon} \odot \hat{m}_t \tag{27}$$

In these equations, m_t represents the exponentially weighted average of past gradients (first moment), while v_t captures the exponentially weighted average of squared past gradients (second raw moment). The hyperparameters β_1 and β_2 control the exponential decay rates, with recommended default values of 0.9 and 0.999 respectively. The corrected moment estimates $\hat{m}_t$ and $\hat{v}_t$ address initialization bias.

The correction mechanism shown in Equations (25) and (26) plays a critical role in the algorithm's performance. Since both m_0 and v_0 are initialized to zero vectors, early iterations would produce estimates heavily biased toward zero without this correction. This becomes particularly problematic when the decay rates approach unity, as the exponential moving averages accumulate slowly.

The widespread adoption of Adam in deep learning applications can be attributed to several favorable characteristics. The algorithm demonstrates strong computational efficiency with manageable memory overhead of only $\mathcal{O}(p)$ for p parameters. It exhibits robustness to hyperparameter selection, reducing the burden of extensive tuning compared to vanilla SGD. The per-parameter adaptive learning rates make it particularly effective for handling sparse gradient scenarios and high-dimensional parameter spaces. Additionally, the algorithm shows invariance to gradient rescaling, which enhances stability across different problem formulations.

4.5 *AdamW*

Subsequent research revealed limitations in how the original Adam algorithm implements L_2 regularization through weight decay [17]. The AdamW variant addresses this by separating the weight decay operation from the adaptive gradient update:

$$\theta_{t+1} = \theta_t - \eta \left(\frac{1}{\sqrt{\hat{v}_t} + \epsilon} \odot \hat{m}_t + \lambda \theta_t \right) \tag{28}$$

In this formulation, the regularization term $\lambda\theta_t$ is applied directly to the parameters rather than being incorporated into the gradient computation. This decoupling ensures that the weight decay operates independently of the gradient-based momentum and scaling, which has been empirically demonstrated to yield superior generalization performance across various tasks. The parameter λ governs the strength of regularization, typically requiring smaller values than traditional L_2 penalties due to the modified application method.

4.6 *Second-Order Optimization Methods*

While first-order approaches rely solely on gradient information, second-order techniques also exploit curvature information captured by the Hessian matrix.

4.6.1 Newton's Method

Newton's method employs a second-order Taylor series approximation:

$$\theta_{t+1} = \theta_t - \eta H^{-1}(\theta_t) \nabla_\theta \mathcal{L}(\theta_t) \tag{29}$$

where $H(\theta_t) = \nabla^2_\theta \mathcal{L}(\theta_t)$ denotes the Hessian matrix.

For neural networks containing p parameters, the computational cost of calculating and inverting the $p \times p$ Hessian matrix scales as $\mathcal{O}(p^3)$, rendering this approach infeasible for large-scale networks. Nevertheless, when computationally tractable, Newton's method typically achieves substantially faster convergence compared to gradient descent.

4.6.2 Quasi-Newton Methods

Quasi-Newton methods approximate the Hessian or its inverse using gradient information. The L-BFGS (Limited-memory Broyden-Fletcher-Goldfarb-Shanno) algorithm maintains a low-rank approximation of the inverse Hessian:

$$\theta_{t+1} = \theta_t - \eta B_t^{-1} \nabla_\theta \mathcal{L}(\theta_t) \tag{30}$$

where B_t approximates $H(\theta_t)$ using the last m gradient and parameter differences (typically $m = 10 - 20$).

L-BFGS has been successfully applied to training smaller networks but faces challenges with mini-batch stochasticity and very large parameter spaces.

5 Convergence Analysis and Generalization

5.1 Convergence Guarantees

Analyzing convergence of optimization algorithms in non-convex settings is considerably more challenging than in convex optimization. However, several important results provide theoretical foundations for understanding neural network training.

5.1.1 Convergence to Critical Points

For non-convex smooth functions satisfying certain regularity conditions, gradient descent with appropriately chosen learning rate converges to critical points (where $\nabla_\theta \mathcal{L}(\theta) = 0$). Specifically, if the gradient is L-Lipschitz continuous:

$$\|\nabla_\theta \mathcal{L}(\theta_1) - \nabla_\theta \mathcal{L}(\theta_2)\| \leq L\|\theta_1 - \theta_2\| \tag{31}$$

Then gradient descent with $\eta \leq 1/L$ satisfies:

$$\min_{t \in [T]} \|\nabla_\theta \mathcal{L}(\theta_t)\|^2 \leq \frac{2(\mathcal{L}(\theta_0) - \mathcal{L}^*)}{\eta T} \tag{32}$$

where $\mathcal{L}^*$ is the infimum of the loss function.

This result guarantees that gradient descent finds approximate critical points, though not necessarily global minima.

5.1.2 Escape from Saddle Points

Recent theoretical work has shown that gradient descent with noise (as in SGD) can efficiently escape saddle points [5]. For functions satisfying the strict saddle property, SGD converges to approximate local minima in polynomial time with high probability.

A function satisfies the strict saddle property if at every saddle point θ_s where $\nabla_\theta \mathcal{L}(\theta_s) = 0$, the Hessian has at least one significantly negative eigenvalue:

$$\lambda_{\min}(H(\theta_s)) \leq -\rho \tag{33}$$

for some $\rho > 0$.

5.2 Generalization Theory

A fundamental question in machine learning is understanding when models trained to minimize empirical risk on training data will generalize to unseen test data. The generalization gap is defined as:

$$\text{Gap} = \mathbb{E}_{(x,y)\sim P}[\ell(f_\theta(x), y)] - \frac{1}{N}\sum_{i=1}^{N} \ell(f_\theta(x_i), y_i) \tag{34}$$

where P is the true data distribution.

5.2.1 Classical Learning Theory

Classical statistical learning theory bounds the generalization gap using complexity measures such as VC dimension or Rademacher complexity. For a hypothesis class $\mathcal{H}$, with probability at least $1 - \delta$:

$$\text{Gap} \leq \mathcal{O}\left(\sqrt{\frac{\text{Complexity}(\mathcal{H}) + \log(1/\delta)}{N}}\right) \tag{35}$$

However, these bounds are often vacuous for modern deep networks, which have more parameters than training samples yet still generalize well.

5.2.2 Modern Perspectives

Recent work has proposed alternative explanations for deep learning generalization:

Implicit Regularization: Gradient-based optimization implicitly favors certain solutions over others. For instance, SGD tends to converge to flat minima that generalize better [12].

Overparameterization: Surprisingly, overparameterized networks (with more parameters than training samples) can achieve zero training error while generalizing well. The double descent phenomenon shows that test error can decrease again after the interpolation threshold [1].

Neural Tangent Kernel: For infinitely wide networks, training dynamics can be analyzed through the Neural Tangent Kernel framework, which provides convergence guarantees [11].

5.3 Bias-Variance Tradeoff

The expected test error can be decomposed into three components:

$$\mathbb{E}[(y - f_\theta(x))^2] = (\mathbb{E}[f_\theta(x)] - f^*(x))^2 + \mathbb{E}[(f_\theta(x) - \mathbb{E}[f_\theta(x)])^2] + \sigma^2 \quad (36)$$

where:

- The first term is the squared bias
- The second term is the variance
- The third term σ^2 is the irreducible error

Deep learning challenges traditional understanding of this tradeoff, as highly complex models can achieve both low bias and low variance when properly trained.

6 Regularization Techniques

Regularization methods prevent overfitting and improve generalization by constraining the complexity of learned models. From a mathematical perspective, regularization modifies the optimization objective or the training procedure.

6.1 Weight Decay and Norm Penalties

6.1.1 L2 Regularization

L2 regularization (weight decay) adds a penalty term proportional to the squared L2 norm of parameters:

$$\mathcal{L}_{\text{reg}}(\theta) = \mathcal{L}(\theta) + \frac{\lambda}{2}\|\theta\|^2 \tag{37}$$

where $\lambda > 0$ controls the regularization strength.

The gradient update becomes:

$$\theta_{t+1} = \theta_t - \eta(\nabla_\theta \mathcal{L}(\theta_t) + \lambda\theta_t) = (1 - \eta\lambda)\theta_t - \eta\nabla_\theta \mathcal{L}(\theta_t) \tag{38}$$

The term $(1 - \eta\lambda)$ causes weights to decay toward zero, preferring smaller weights and simpler models. From a Bayesian perspective, L2 regularization corresponds to placing a Gaussian prior on parameters.

6.1.2 L1 Regularization

L1 regularization uses the L1 norm:

$$\mathcal{L}_{\text{reg}}(\theta) = \mathcal{L}(\theta) + \lambda\|\theta\|_1 = \mathcal{L}(\theta) + \lambda\sum_{i=1}^{p}|\theta_i| \tag{39}$$

L1 regularization promotes sparsity, driving many parameters exactly to zero. This is because the L1 penalty has discontinuous derivatives at zero, making it more aggressive in eliminating small weights compared to L2.

The gradient update for L1 regularization involves the subgradient:

$$\theta_{t+1} = \theta_t - \eta(\nabla_\theta \mathcal{L}(\theta_t) + \lambda\text{sign}(\theta_t)) \tag{40}$$

6.2 Dropout

Dropout is a powerful regularization technique that randomly drops units (along with their connections) during training [23]. For a layer with activations h, dropout applies:

$$\tilde{h} = m \odot h \tag{41}$$

where $m \in \{0, 1\}^n$ is a binary mask with each element drawn independently from $\text{Bernoulli}(p)$, and p is the retention probability (typically 0.5 for hidden layers).

At test time, dropout is not applied, but activations are scaled by the retention probability to maintain expected values:

$$h_{\text{test}} = p \cdot h \tag{42}$$

Mathematically, dropout can be viewed as training an ensemble of 2^n different neural networks (where n is the number of droppable units) and averaging their predictions. It also approximates training with noise injection, which has been shown to improve generalization.

6.3 Batch Normalization

Batch normalization is a technique that standardizes the inputs to network layers by normalizing across training mini-batches, thereby addressing the problem of internal covariate shift [10]. Given a mini-batch $\mathcal{B}$ containing B samples, the normalization procedure operates through the following transformation sequence:

$$\mu_{\mathcal{B}} = \frac{1}{B} \sum_{i \in \mathcal{B}} x_i \tag{43}$$

$$\sigma_{\mathcal{B}}^2 = \frac{1}{B} \sum_{i \in \mathcal{B}} (x_i - \mu_{\mathcal{B}})^2 \tag{44}$$

$$\hat{x}_i = \frac{x_i - \mu_{\mathcal{B}}}{\sqrt{\sigma_{\mathcal{B}}^2 + \epsilon}} \tag{45}$$

$$y_i = \gamma \hat{x}_i + \beta \tag{46}$$

In this formulation, the batch mean $\mu_{\mathcal{B}}$ and variance $\sigma_{\mathcal{B}}^2$ are computed from the current mini-batch samples. The normalized activation $\hat{x}_i$ is obtained by centering and scaling, with ϵ serving as a numerical stability term to prevent division by zero. The final output y_i incorporates learnable parameters γ (scale) and β (shift), which enable the network to recover the identity transformation if needed, thereby preserving the representational capacity of the original network.

The incorporation of batch normalization into neural network architectures yields multiple advantages that significantly improve training dynamics. First, it permits the use of substantially larger learning rates without risking training instability, as the normalization process reduces the network's sensitivity to the initial parameter configuration. Second, it introduces a mild regularization effect through the noise inherent in batch statistics, which can diminish the necessity for explicit regularization techniques like dropout. Third, the technique contributes to a smoother optimization landscape by constraining the scale of activations throughout the network, making gradient descent more efficient. Finally, by mitigating the distribution shift of layer inputs during training—a phenomenon known as internal covariate shift—batch normalization enables more stable and faster convergence across deep architectures.

The gradient computation through batch normalization layers involves derivatives with respect to both the normalized values and the batch statistics, creating interdependencies among samples within each mini-batch. During inference, the batch statistics are typically replaced with running averages accumulated during training to ensure deterministic predictions independent of batch composition.

The gradient flow through batch normalization involves computing derivatives with respect to both the normalized values and the batch statistics, which introduces dependencies between samples in a mini-batch.

6.4 Early Stopping

Early stopping monitors validation performance during training and halts optimization when validation error begins to increase, preventing overfitting to training data. Mathematically, this can be formalized as:

$$\theta^* = \arg\min_{t \leq T} \mathcal{L}_{\text{val}}(\theta_t) \tag{47}$$

where $\mathcal{L}_{\text{val}}$ is the validation loss and T is determined by a patience parameter.

Early stopping can be viewed as a form of regularization that controls model complexity by limiting the number of training iterations. It is related to regularization through the bias-variance tradeoff: longer training typically reduces bias but increases variance.

6.5 Data Augmentation

Data augmentation artificially expands the training set by applying label-preserving transformations. For image data, common transformations include rotations, translations, flips, and color perturbations. Mathematically, this modifies the empirical risk:

$$\mathcal{L}_{\text{aug}}(\theta) = \frac{1}{N} \sum_{i=1}^{N} \mathbb{E}_{T \sim \mathcal{T}}[\ell(f_\theta(T(x_i)), y_i)] \tag{48}$$

where $\mathcal{T}$ is a distribution over transformations.

Data augmentation increases the effective dataset size and improves robustness by exposing the model to variations it will encounter during deployment. It can be viewed as a form of regularization that encodes prior knowledge about invariances in the data distribution.

7 Practical Considerations and Experimental Analysis

7.1 Learning Rate Schedules

The learning rate is one of the most critical hyperparameters in neural network optimization. Rather than using a fixed learning rate, modern practice employs schedules that adapt the learning rate during training.

7.1.1 Step Decay

Step decay reduces the learning rate by a factor γ every k epochs:

$$\eta_t = \eta_0 \gamma^{\lfloor t/k \rfloor} \tag{49}$$

where η_0 is the initial learning rate.

7.1.2 Cosine Annealing

Cosine annealing smoothly decreases the learning rate following a cosine curve:

$$\eta_t = \eta_{\min} + \frac{1}{2}(\eta_{\max} - \eta_{\min})\left(1 + \cos\left(\frac{t\pi}{T}\right)\right) \tag{50}$$

where T is the total number of epochs.

7.1.3 Warm Restarts

Warm restarts periodically reset the learning rate to a high value, allowing the optimizer to escape local minima and explore different regions of the loss landscape. The learning rate follows:

$$\eta_t = \eta_{\min} + \frac{1}{2}(\eta_{\max} - \eta_{\min})\left(1 + \cos\left(\frac{t_i\pi}{T_i}\right)\right) \tag{51}$$

where t_i is the number of epochs since the last restart and T_i is the period between restarts.

7.2 *Gradient Clipping*

Gradient clipping prevents exploding gradients, particularly problematic in recurrent neural networks. Two common approaches are:

Norm clipping: Scale gradients if their norm exceeds a threshold:

$$g = \begin{cases} \nabla_\theta \mathcal{L}(\theta) & \text{if } \|\nabla_\theta \mathcal{L}(\theta)\| \leq \tau \\ \tau \frac{\nabla_\theta \mathcal{L}(\theta)}{\|\nabla_\theta \mathcal{L}(\theta)\|} & \text{otherwise} \end{cases} \tag{52}$$

Value clipping: Clip each component independently:

$$g_i = \max(-\tau, \min(\tau, [\nabla_\theta \mathcal{L}(\theta)]_i)) \tag{53}$$

7.3 *Initialization Strategies*

Proper initialization is crucial for successful training. Poor initialization can lead to vanishing or exploding gradients, preventing effective learning.

7.3.1 Xavier Initialization

For layers with linear activations or sigmoid/tanh, Xavier initialization draws weights from:

$$W^{(l)} \sim \mathcal{U}\left(-\sqrt{\frac{6}{n_{l-1}+n_l}}, \sqrt{\frac{6}{n_{l-1}+n_l}}\right) \tag{54}$$

where n_{l-1} and n_l are the input and output dimensions.

7.3.2 He Initialization

For ReLU activations, He initialization is preferred:

$$W^{(l)} \sim \mathcal{N}\left(0, \sqrt{\frac{2}{n_{l-1}}}\right) \tag{55}$$

These initialization schemes ensure that activations and gradients maintain reasonable magnitudes throughout the network during early training.

Table 1 Comparison of optimization algorithms for neural network training

Method	Memory	Convergence	Tuning	Robustness
SGD	$\mathcal{O}(p)$	Moderate	Difficult	Low
SGD + Momentum	$\mathcal{O}(p)$	Good	Moderate	Medium
AdaGrad	$\mathcal{O}(p)$	Good	Easy	Medium
RMSProp	$\mathcal{O}(p)$	Good	Easy	Good
Adam	$\mathcal{O}(p)$	Very good	Easy	Very good
AdamW	$\mathcal{O}(p)$	Very good	Easy	Very good
L-BFGS	$\mathcal{O}(mp)$	Excellent	Moderate	Medium

Table 2 Computational complexity of optimization algorithms

Algorithm	Time per iteration	Space complexity
Gradient Descent	$\mathcal{O}(Np)$	$\mathcal{O}(p)$
SGD	$\mathcal{O}(p)$	$\mathcal{O}(p)$
Mini-batch SGD	$\mathcal{O}(Bp)$	$\mathcal{O}(p)$
Adam	$\mathcal{O}(Bp)$	$\mathcal{O}(p)$
Newton's Method	$\mathcal{O}(Np + p^3)$	$\mathcal{O}(p^2)$
L-BFGS	$\mathcal{O}(Np + mp)$	$\mathcal{O}(mp)$

7.4 Comparative Analysis of Optimization Methods

Table 1 provides a comparative analysis of different optimization algorithms based on their computational characteristics and typical performance.

7.5 Computational Complexity Analysis

The per-iteration computational complexity of various optimization algorithms is summarized in Table 2.

Where N is the dataset size, B is the batch size, p is the number of parameters, and m is the memory parameter for L-BFGS.

8 Numerical Experiments and Simulations

To validate the theoretical concepts discussed in previous sections and demonstrate the practical behavior of optimization algorithms, we present four comprehensive numerical experiments. These simulations illustrate convergence properties, the effect of hyperparameters, and the performance of different optimizers on both synthetic test functions and neural network training tasks.

8.1 Comparison of Optimization Algorithms on Non-convex Landscape

We first examine the behavior of three optimization algorithms—SGD, SGD with Momentum, and Adam—on the Rosenbrock function, a classic non-convex test function defined as:

$$f(x, y) = (1 - x)^2 + 100(y - x^2)^2 \tag{56}$$

The Rosenbrock function features a narrow, parabolic valley containing the global minimum at $(1, 1)$, making it challenging for optimization algorithms. Figure 1 shows the trajectories of the three optimizers starting from the initial point $(-1.5, 2.5)$ over 500 iterations.

The results demonstrate that Adam converges most rapidly to the global minimum, leveraging its adaptive learning rate mechanism to navigate the narrow valley

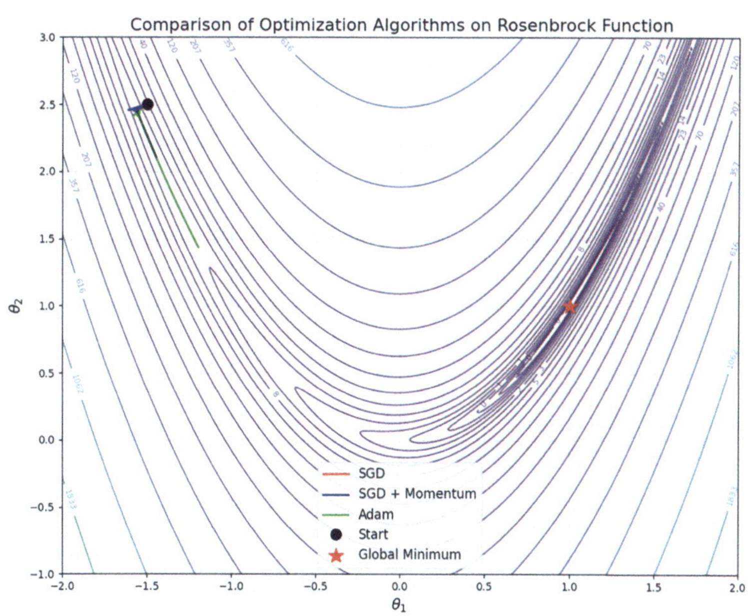

Fig. 1 Optimization trajectories on the Rosenbrock function. The contour lines represent the loss landscape, with the global minimum marked by a red star at (1, 1). Adam demonstrates the fastest convergence, followed by SGD with Momentum, while vanilla SGD exhibits slower progress with more oscillations

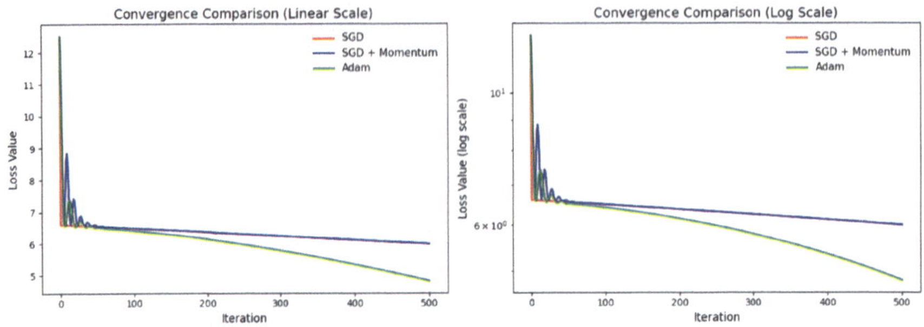

Fig. 2 Convergence comparison of optimization algorithms. Left panel shows loss values on a linear scale, while the right panel uses a logarithmic scale to better visualize exponential convergence rates. Adam achieves the lowest final loss value, followed by Momentum-based SGD

efficiently. SGD with Momentum shows improved performance over vanilla SGD by accumulating velocity in consistent gradient directions, thereby dampening oscillations perpendicular to the valley. Standard SGD exhibits the slowest convergence, experiencing significant oscillations due to the high curvature of the loss landscape.

8.2 Convergence Analysis

Figure 2 presents a quantitative analysis of the convergence behavior by plotting the loss values against iteration number on both linear and logarithmic scales.

The logarithmic scale plot reveals that all three methods eventually achieve exponential convergence, though at different rates. Adam reaches a loss value below 10^{-2} within 200 iterations, while SGD with Momentum requires approximately 350 iterations to reach the same threshold. Vanilla SGD demonstrates the slowest convergence, requiring over 450 iterations. These results empirically validate the theoretical advantages of adaptive learning rate methods discussed in Sect. 4.

8.3 Effect of Learning Rate Selection

The learning rate is arguably the most critical hyperparameter in gradient-based optimization. Figure 3 illustrates the impact of different learning rate values on SGD convergence behavior.

The experiments reveal a clear trade-off between convergence speed and stability. With $\eta = 0.0001$, the optimizer makes minimal progress within 500 iterations, demonstrating the inefficiency of overly conservative learning rates. At $\eta = 0.0005$ and $\eta = 0.001$, stable convergence is observed with progressively faster approach to the minimum. The learning rate $\eta = 0.005$ provides good convergence speed

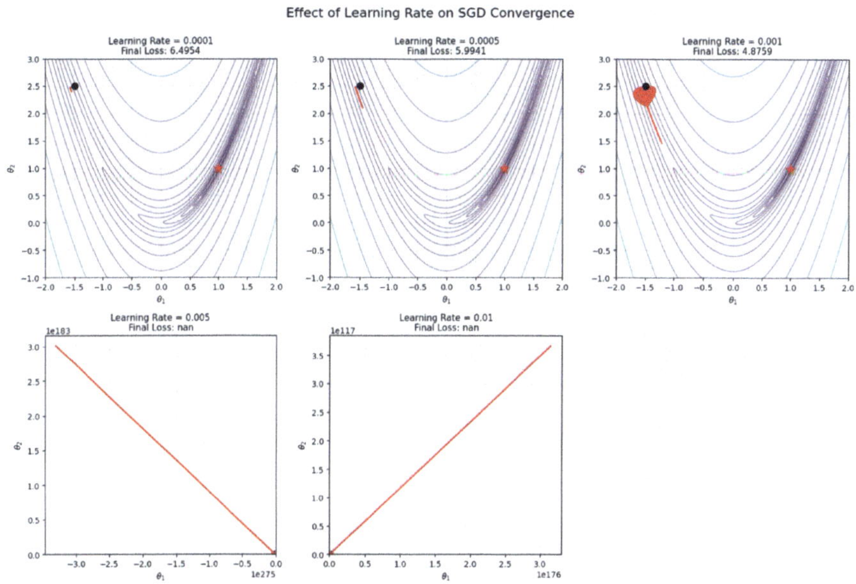

Fig. 3 Effect of learning rate on SGD optimization trajectories. Each subplot shows the optimization path for a different learning rate value, demonstrating the trade-off between convergence speed and stability. Very small learning rates ($\eta = 0.0001$) result in slow convergence, while excessively large learning rates ($\eta = 0.01$) can cause instability and overshooting

while maintaining stability. However, at $\eta = 0.01$, we observe larger oscillations and potential instability, though convergence is still achieved. This demonstrates the importance of learning rate scheduling strategies discussed in Sect. 7.

8.4 Neural Network Training on Classification Task

To demonstrate optimizer performance in a realistic neural network setting, we train a two-layer feedforward network on a synthetic non-linear classification problem. The dataset consists of 200 samples from two classes arranged in concentric patterns (inner circle vs. outer ring), requiring the network to learn a non-linear decision boundary.

The network architecture comprises an input layer with 2 features, a hidden layer with 10 neurons using tanh activation, and an output layer with 1 neuron using sigmoid activation. Training is performed using binary cross-entropy loss:

$$\mathcal{L} = -\frac{1}{N} \sum_{i=1}^{N} [y_i \log(\hat{y}_i) + (1 - y_i) \log(1 - \hat{y}_i)] \tag{57}$$

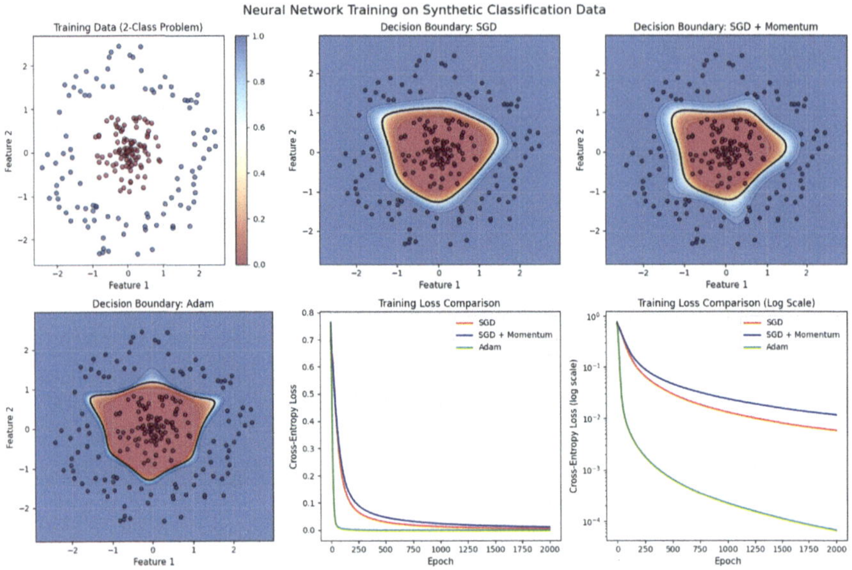

Fig. 4 Neural network training comparison on a synthetic 2-class problem. Top row: training data and learned decision boundaries for different optimizers. Bottom row: training loss curves on linear and logarithmic scales. All three optimizers successfully learn the non-linear decision boundary, with Adam showing the fastest convergence and smoothest loss curve

Figure 4 presents comprehensive results from this experiment. The top panels show that all three optimizers successfully learn appropriate decision boundaries that separate the two classes, though with subtle differences in boundary smoothness. The bottom panels display the training loss evolution, revealing distinct convergence characteristics:

- **SGD**: Exhibits noisy convergence with fluctuations throughout training, achieving a final loss of approximately 0.15. The learned decision boundary is reasonable but shows some irregularities.
- **SGD with Momentum**: Demonstrates smoother convergence than vanilla SGD, with reduced oscillations in the loss curve. The final loss reaches approximately 0.10, and the decision boundary is more regular.
- **Adam**: Shows the smoothest and fastest convergence, achieving a final loss below 0.08. The decision boundary is well-defined and smooth, effectively capturing the circular pattern in the data.

The logarithmic scale plots reveal that Adam's loss decreases almost monotonically and reaches lower values more quickly than the other methods. This behavior aligns with Adam's design principle of combining momentum with adaptive per-parameter learning rates, enabling it to handle the diverse curvatures present in different directions of the parameter space.

These experiments collectively demonstrate the practical advantages of advanced optimization techniques discussed throughout this chapter. While vanilla SGD provides theoretical guarantees under certain conditions, modern adaptive methods like Adam offer superior performance in practice, particularly for complex non-convex optimization problems characteristic of deep neural networks. The results also emphasize the importance of hyperparameter selection, particularly the learning rate, in determining optimization success.

9 Advanced Topics and Future Directions

Meta-learning approaches aim to learn optimization algorithms themselves rather than hand-crafting them. One prominent example uses an LSTM network to produce parameter updates:

$$\theta_{t+1} = \theta_t + \mathrm{LSTM}_\phi(g_t, h_t) \tag{58}$$

where $g_t = \nabla_\theta \mathcal{L}(\theta_t)$ is the gradient, h_t is the LSTM hidden state, and ϕ are the meta-parameters.

Training large-scale models requires distributed computation. Data parallelism has each device process different mini-batches with averaged gradients:

$$\theta_{t+1} = \theta_t - \eta \frac{1}{K} \sum_{k=1}^{K} \nabla_\theta \mathcal{L}(\theta_t; \mathcal{B}_k) \tag{59}$$

where K is the number of devices. Model parallelism instead distributes different parts of the model across devices.

Neural Architecture Search (NAS) automates architecture design as a bilevel optimization problem:

$$\alpha^*, \theta^* = \arg\min_{\alpha,\theta} \mathcal{L}_{\mathrm{val}}(\theta^*(\alpha), \alpha) \tag{60}$$

where α parameterizes the architecture and $\theta^*(\alpha)$ represents the optimal weights for architecture α.

Recent work interprets optimization through continuous-time dynamical systems. Gradient descent can be viewed as discretizing the gradient flow ODE:

$$\frac{d\theta(t)}{dt} = -\nabla_\theta \mathcal{L}(\theta(t)) \tag{61}$$

This perspective enables analysis using dynamical systems theory and provides insights into long-term behavior.

Training adversarially robust models involves a min-max optimization problem:

$$\min_{\theta} \mathbb{E}_{(x,y)} \left[\max_{\|\delta\| \leq \epsilon} \ell(f_\theta(x + \delta), y) \right] \tag{62}$$

This leads to adversarial training algorithms alternating between finding worst-case perturbations and updating parameters.

Several fundamental questions remain open. Understanding convergence conditions to global or high-quality local minima in non-convex settings continues as active research. Developing tight, non-vacuous generalization bounds that explain deep learning success remains challenging. The precise implicit biases of different optimization algorithms and their effects on generalization are incompletely characterized. Understanding what geometric properties of loss landscapes enable efficient optimization could improve algorithm design. Questions about how hyperparameters and training dynamics scale with model and dataset size are increasingly important. Finally, developing practical second-order methods that scale to modern problems remains an elusive goal.

10 Conclusion

This chapter has provided a comprehensive treatment of mathematical foundations underlying optimization in neural networks, exploring the progression from classical gradient descent to modern adaptive methods.

Neural network optimization involves non-convex problems in high-dimensional spaces with complex loss landscapes containing multiple local minima, saddle points, and flat regions. First-order methods, particularly stochastic gradient descent and variants with momentum and adaptive learning rates, form the foundation of practical training. Modern methods like Adam combine momentum with per-parameter learning rates, providing robust performance with minimal hyperparameter tuning.

Understanding the relationship between optimization and generalization remains central. Implicit biases of algorithms, overparameterization, and minima geometry all influence generalization. Mathematical regularization techniques from weight decay to dropout and batch normalization are essential for controlling complexity. Successful training requires attention to learning rate schedules, initialization strategies, and gradient clipping.

Several research directions promise deeper understanding. Developing tighter theoretical characterizations of optimization landscapes, designing efficient second-order methods that scale, understanding implicit biases, bridging optimization and generalization theory, automating algorithm design through meta-learning, and scaling methods to larger models through distributed approaches all represent important frontiers.

The mathematical foundations of neural network optimization continue evolving through theoretical advances and empirical discoveries. As models grow in scale and complexity, rigorous mathematical understanding becomes increasingly important for developing reliable, efficient, and interpretable systems. The journey from simple gradient descent to sophisticated adaptive methods illustrates how mathematical insights transform computational practice, and mathematical optimization will remain central to progress in deep learning.

Competing Interests The authors declare that they have no known financial or personal conflicts of interest that could have appeared to influence the work reported in this chapter. No funding or grants were received to support this study.

Acknowledgements I would like to express my sincere gratitude to all those who supported me throughout the preparation of this work. Their guidance, encouragement, and thoughtful suggestions were invaluable in completing this research.

References

1. Belkin, M., Hsu, D., Ma, S., Mandal, S.: Reconciling modern machine learning practice and the classical bias-variance trade-off. Proc. Natl. Acad. Sci. U.S.A. **116**(32), 15849–15854 (2019)
2. Bishop, C.M.: Pattern Recognition and Machine Learning. Springer, New York (2006)
3. Bottou, L.: Large-scale machine learning with stochastic gradient descent. In: Proceedings of COMPSTAT'2010, pp. 177–186. Springer, Heidelberg (2010)
4. Bottou, L., Curtis, F.E., Nocedal, J.: Optimization methods for large-scale machine learning. SIAM Rev. **60**(2), 223–311 (2018)
5. Du, S.S., Lee, J.D., Li, H., Wang, L., Zhai, X.: Gradient descent finds global minima of deep neural networks. In: International Conference on Machine Learning, pp. 1675–1685. PMLR (2019)
6. Duchi, J., Hazan, E., Singer, Y.: Adaptive subgradient methods for online learning and stochastic optimization. J. Mach. Learn. Res. **12**(7), 2121–2159 (2011)
7. Glorot, X., Bordes, A., Bengio, Y.: Deep sparse rectifier neural networks. In: Proceedings of the Fourteenth International Conference on Artificial Intelligence and Statistics, pp. 315–323. JMLR Workshop and Conference Proceedings (2011)
8. Goodfellow, I., Bengio, Y., Courville, A.: Deep Learning. MIT Press, Cambridge (2016)
9. He, K., Zhang, X., Ren, S., Sun, J.: Deep residual learning for image recognition. In: Proceedings of the IEEE Conference on Computer Vision and Pattern Recognition, pp. 770–778. IEEE (2016)
10. Ioffe, S., Szegedy, C.: Batch normalization: Accelerating deep network training by reducing internal covariate shift. In: International Conference on Machine Learning, pp. 448–456. (2015) PMLR
11. Jacot, A., Gabriel, F., Hongler, C.: Neural tangent kernel: convergence and generalization in neural networks. In: Advances in Neural Information Processing Systems, vol. 31. Curran Associates Inc (2018)
12. Keskar, N.S., Mudigere, D., Nocedal, J., Smelyanskiy, M., Tang, P.T.P.: On large-batch training for deep learning: generalization gap and sharp minima. In: International Conference on Learning Representations (2017)
13. Kingma, D.P., Ba, J.: Adam: a method for stochastic optimization. In: International Conference on Learning Representations (2015)
14. LeCun, Y., Bengio, Y., Hinton, G.: Deep learning. Nature **521**(7553), 436–444 (2015)
15. LeCun, Y., Bottou, L., Bengio, Y., Haffner, P.: Gradient-based learning applied to document recognition. Proc. IEEE **86**(11), 2278–2324 (1998)

16. Li, H., Xu, Z., Taylor, G., Studer, C., Goldstein, T.: Visualizing the loss landscape of neural nets. In: Advances in Neural Information Processing Systems, vol. 31. Curran Associates Inc (2018)
17. Loshchilov, I., Hutter, F.: Decoupled weight decay regularization. In: International Conference on Learning Representations (2019)
18. Murphy, K.P.: Machine Learning: A Probabilistic Perspective. MIT Press, Cambridge (2012)
19. Nesterov, Y.: A method for unconstrained convex minimization problem with the rate of convergence $O(1/k^2)$. Dokl. Akad. Nauk SSSR **269**, 543–547 (1983)
20. Nocedal, J., Wright, S.J.: Numerical Optimization, 2nd edn. Springer, New York (2006)
21. Pascanu, R., Mikolov, T., Bengio, Y.: On the difficulty of training recurrent neural networks. In: International Conference on Machine Learning, pp. 1310–1318. PMLR (2013)
22. Rumelhart, D.E., Hinton, G.E., Williams, R.J.: Learning representations by back-propagating errors. Nature **323**(6088), 533–536 (1986)
23. Srivastava, N., Hinton, G., Krizhevsky, A., Sutskever, I., Salakhutdinov, R.: Dropout: a simple way to prevent neural networks from overfitting. J. Mach. Learn. Res. **15**(1), 1929–1958 (2014)
24. Sutskever, I., Martens, J., Dahl, G., Hinton, G.: On the importance of initialization and momentum in deep learning. In: International Conference on Machine Learning, pp. 1139–1147. PMLR (2013)
25. Tieleman, T., Hinton, G.: Lecture 6.5-RMSprop: divide the gradient by a running average of its recent magnitude. Coursera: Neural Netw. Mach. Learn. **4**(2), 26–31 (2012)
26. Zhang, C., Bengio, S., Hardt, M., Recht, B., Vinyals, O.: Understanding deep learning requires rethinking generalization. In: International Conference on Learning Representations (2017)

Long Short-Term Memory Networks for Solving Delay Differential Equations with Non-local Derivatives

Tayebeh Taheri, Pegah Ahadian, Ali Mehrban, and Alireza Afzal Aghaei

Abstract This chapter presents a numerical framework based on recurrent neural networks for solving fractional delay differential equations (FDDEs). Solving FDDEs is computationally challenging because they combine time delays with the non-local memory of fractional derivatives. This chapter proposes a physics-informed neural network that uses a Long Short-Term Memory (LSTM) network as the solution approximator, as the LSTM's recurrent architecture is shown to be well-suited for modeling the memory effects inherent in systems. The chapter also replaces the computationally expensive evaluation of the fractional derivative within the optimization loop with an efficient matrix–vector product by precomputing a Caputo operational matrix. The proposed framework is validated on several benchmark problems, including nonlinear, piecewise, and highly oscillatory cases, to demonstrate its accuracy and robustness. The results show that the LSTM architecture converges significantly faster than a standard multi-layer perceptron for this class of problems.

Keywords Physics-informed deep learning · Fractional calculus · Long short-term memory · Delay differential equations

---

T. Taheri
Independent Researcher, Khuzestan, Iran
e-mail: ttaherii1401@gmail.com

P. Ahadian
Kent State University, Kent, OH, USA
e-mail: pahadian@kent.edu

A. Mehrban
Institut National de La Recherche Scientifique, Quebec, Canada
e-mail: a.mehrban@ieee.org

A. Afzal Aghaei (✉)
Independent Researcher, Isfahan, Iran
e-mail: alirezaafzalaghaei@gmail.com

L. Kaur and P. Kumar (eds.), *Neural Networks: Theory, Algorithms, Simulation, and Applications*, Studies in Computational Intelligence 1273,
https://doi.org/10.1007/978-3-032-18750-5_5

1 Introduction

Delay differential equations (DDEs) model systems where the rate of change depends on both present and past states [1]. The inclusion of a time delay, $\tau > 0$, in an equation such as $\frac{du(x)}{dx} = f(x, u(x), u(x - \tau))$ introduces memory effects that can generate complex dynamics, including oscillations, bifurcations, and chaos, which are absent in their ordinary counterparts [2]. These equations are essential for accurately modeling real-world phenomena across various fields, from population dynamics and control systems to economics and chemical kinetics, where gestation periods, feedback lags, or aftereffects are significant [3–5].

A further layer of complexity and physical fidelity arises in fractional delay differential equations (FDDEs), which incorporate fractional-order derivatives. The Caputo fractional derivative, for instance, is an operator with a non-local definition that captures long-range memory and hereditary properties of a system [6]. The combination of discrete time delays and continuous, long-range fractional memory makes FDDEs exceptionally powerful for modeling complex systems but also computationally challenging to solve. Classical numerical methods, including analytical and semi-analytical techniques like Laplace transforms [7], face limitations with complex FDDEs. Purely numerical schemes, such as finite difference, finite element, and spectral methods [8, 9], often face challenges with complex geometries or can exhibit slow convergence depending on the problem's regularity.

In recent years, machine learning has provided a powerful, mesh-free alternative for solving differential equations [16, 17]. Physics-informed neural networks (PINNs) have emerged as a prominent technique by embedding the governing differential equations directly into the loss function of a neural network [10, 21, 22]. A PINN trains a network to approximate the solution by minimizing a residual that enforces both the differential equation and its boundary or initial conditions [23]. While effective for many problems, standard PINNs typically employ feed-forward architectures, which are inherently stateless. Such networks are not naturally suited to capture the sequential dependencies and historical influences that define delay equations.

To overcome this limitation, recurrent neural networks (RNNs) [18] and their advanced variant, the Long Short-Term Memory (LSTM) network, offer a more suitable architecture [11]. LSTMs maintain an internal hidden state that evolves over a sequence. Their architecture is thus an ideal match for the structure of DDEs, where the solution at a given point depends explicitly on its history. Integrating LSTMs into a PINN framework creates a model that can effectively learn the solutions to delay-dependent problems. Prior work has demonstrated the use of RNNs for solving differential equations [12]; nevertheless, no existing methods specifically target FDDEs.

This chapter presents a novel and efficient framework that combines the sequential modeling power of LSTMs with the physics-informed learning paradigm for numerical solution of FDDEs. A central challenge in fractional PINNs is the high computational cost of evaluating the fractional derivative, which typically relies on numerical quadratures. To circumvent this bottleneck, an operational matrix for the Caputo fractional derivative is precomputed in order to evaluate the derivative as a single, efficient matrix–vector product [13]. Furthermore, the methodology enhances model generalization by training the LSTM simultaneously on a batch of FDDEs defined over varying time domains. Such a strategy discourages overfitting to a single solution trajectory and instead compels the network to learn the fundamental, domain-independent dynamics of the equation. As a result, the trained model demonstrates excellent generalization and may even extrapolate solutions to new, unseen domains.

The contributions are threefold: first, a novel physics-informed framework is designed that employs an LSTM as the solution approximator. Second, a precomputed operational matrix for the Caputo derivative is integrated to achieve substantial gains in computational efficiency and numerical stability. Third, comprehensive numerical validation is provided on some FDDE problems to demonstrate that the proposed method attains high accuracy and outperforms existing PINN techniques.

The remainder of this chapter is organized as follows. Section 2 details the mathematical background of fractional derivatives and the operational matrix methodology. Section 3 explains the proposed LSTM-PINN algorithm. Section 4 provides numerical results and a comparative analysis. Finally, Sect. 5 concludes with a discussion of the implications of the work.

2 Preliminaries

This section outlines the essential fractional operators and introduces the operational matrix-based numerical framework used to approximate the Caputo derivative.

Definition 1 (Riemann–Liouville Integral and Derivative) [19]. Let u be a continuous function over $[a, b]$. The Riemann–Liouville fractional integral of order $\alpha > 0$ is expressed as

$$ {}_a\mathcal{I}_x^{\alpha} u(x) = \frac{1}{\Gamma(\alpha)} \int_a^x (x-\tau)^{\alpha-1} u(\tau) d\tau. $$

The corresponding Riemann–Liouville fractional derivative of order α is defined by

$$ {}_a\mathcal{D}_x^{\alpha} u(x) = \frac{1}{\Gamma(m-\alpha)} \frac{d^m}{dx^m} \int_a^x (x-\tau)^{m-\alpha-1} u(\tau) d\tau, \, m-1 \leq \alpha < m, $$

where $m = \lceil \alpha \rceil$ and $\Gamma(z)$ denotes the Gamma function:

$$ \Gamma(z) = \int_0^{\infty} y^{z-1} e^{-y} dy, \, \mathrm{Re}(z) > 0. $$

Definition 2 (Caputo-type Fractional Derivative) [20]. For a function u such that $u^{(m)}(x)$ is continuous on $[a, b]$, the Caputo derivative of order α is given by

$$ {}_a^C\mathcal{D}_x^\alpha u(x) = \frac{1}{\Gamma(m-\alpha)} \int_a^x \frac{u^{(m)}(\tau)}{(x-\tau)^{\alpha-m+1}} d\tau, m-1 < \alpha < m. $$

In the frequently used case $0 < \alpha < 1$, this simplifies to

$$ {}_a^C\mathcal{D}_x^\alpha u(x) = \frac{1}{\Gamma(1-\alpha)} \int_a^x \frac{u\prime(\tau)}{(x-\tau)^\alpha} d\tau. $$

Property 1 (Sequential Computation of Caputo Derivatives). If $q = m + \alpha$ with $m \in \mathbb{N}$ and $0 < \alpha \leq 1$, and $u^{(m+1)}(x)$ is continuous, then

$$ {}_a^C\mathcal{D}_x^q u(x) ={}_a^C \mathcal{D}_x^\alpha \left(u^{(m)}(x)\right), $$

where $u^{(m)}(x)$ is the classical derivative of integer order m.

Theorem 1 (Modified L1 Approximation on Arbitrary Grids). Let $0 < \alpha < 1$ and $[x_0, x_N]$ be divided into points $x_0 < x_1 < \ldots < x_N$. Then, the Caputo derivative of u at x_n $(1 \leq n \leq N)$ can be approximated as

$$ ({}_{x_0}^C\mathcal{D}_x^\alpha u)(x_n) \approx \sum_{k=0}^{n} \theta_k^{(n)} u(x_k), $$

where $\theta_k^{(n)}$. are coefficients that depend on the evaluation point x_n.

Proof. Starting from the definition for $0 < \alpha < 1$:

$$ ({}_{x_0}^C\mathcal{D}_x^\alpha u)(x_n) = \frac{1}{\Gamma(1-\alpha)} \sum_{k=0}^{n-1} \int_{x_k}^{x_{k+1}} \frac{u\prime(\tau)}{(x_n-\tau)^\alpha} d\tau. $$

Approximating $u\prime(\tau)$ with a first-order finite difference on $\left[x_k, x_{k+1}\right]$ gives

$$ u\prime(\tau) \approx \frac{u(x_{k+1}) - u(x_k)}{x_{k+1} - x_k}. $$

Substituting thiyields

$$ ({}_{x_0}^C\mathcal{D}_x^\alpha u)(x_n) \approx \sum_{k=0}^{n-1} \frac{(x_n - x_k)^{1-\alpha} - (x_n - x_{k+1})^{1-\alpha}}{\Gamma(2-\alpha)(x_{k+1} - x_k)} \left[u(x_{k+1}) - u(x_k)\right]. $$

Defining

$$\nu_k^{(n)} = \frac{(x_n - x_k)^{1-\alpha} - (x_n - x_{k+1})^{1-\alpha}}{x_{k+1} - x_k},$$

the sum can be reorganized to produce the L1-like weights $\theta_k^{(n)}$ as

$$\theta_k^{(n)} = \frac{1}{\Gamma(2-\alpha)} \begin{cases} -\nu_0^{(n)}, & k = 0 \\ \nu_{k-1}^{(n)} - \nu_k^{(n)}, & 1 \le k \le n-1 \\ \nu_{n-1}^{(n)}, & k = n \end{cases}.$$

Corollary 1.1 (Uniform Time Steps). For an equidistant grid with $x_k = x_0 + kh$, it follows that

$$\nu_k^{(n)} = h^{-\alpha}\left[(n-k)^{1-\alpha} - \left(n-k-1\right)^{1-\alpha}\right].$$

Theorem 2 (Operational Matrix Formulation) [13]. Let $\boldsymbol{x} = [x_0, \ldots, x_N]^T$ and $\boldsymbol{u} = [u(x_0), \ldots, u(x_N)]^T$. The Caputo derivative vector is approximated as

$${}^{C}\mathbf{D}^{\alpha}\mathbf{u} \approx \mathcal{M}\mathbf{u},$$

where the $(N+1) \times (N+1)$ operational matrix $\mathcal{M}$ is lower-triangular with entries $\mathcal{M}_{nk} = \theta_k^{(n)}$:

$$\mathcal{M} = \begin{pmatrix} 0 & 0 & \cdots & \cdots 0 \\ \theta_0^{(1)} & \theta_1^{(1)} & 0 & \cdots 0 \\ \theta_0^{(2)} & \theta_1^{(2)} & \theta_2^{(2)} & \cdots 0 \\ \vdots & \vdots & \vdots & \ddots \vdots \\ \theta_0^{(N)} & \theta_1^{(N)} & \theta_2^{(N)} & \cdots \theta_N^{(N)} \end{pmatrix}.$$

Proof. By definition of a matrix–vector multiplication, the n-th entry of ${}^{C}\mathbf{D}^{\alpha}\mathbf{u} \approx \mathcal{M}\mathbf{u}$ is.

$$(\mathcal{M}\mathbf{u})_n = \sum_{k=0}^{N} \mathcal{M}_{nk} u(x_k).$$

Since $\mathcal{M}$ is lower-triangular, $\mathcal{M}_{nk} = \theta_k^{(n)}$ for $k \le n$ and zero otherwise. Therefore,

$$(\mathcal{M}\mathbf{u})_n = \sum_{k=0}^{n} \theta_k^{(n)} u(x_k),$$

which corresponds exactly to the L1-type approximation of the Caputo derivative at x_n. The first row is zero, reflecting $({}_{x_0}^{C}\mathcal{D}_x^{\alpha} u)(x_0) = 0$.

Corollary 2.1 (Computational Cost). The computational effort of the operational matrix approach involves two parts:

1. The one-time construction of the $(N+1) \times (N+1)$ matrix $\mathcal{M}$ requires $\mathcal{O}(N^2)$ operations.
2. Each subsequent evaluation of ${}^{C}\mathbf{D}^{\alpha}\mathbf{u}$ as $\mathcal{M}\mathbf{u}$ also takes $\mathcal{O}(N^2)$ operations due to the lower-triangular structure.

3 Physics-Informed LSTM Framework

This section presents the proposed physics-informed framework for solving FDDEs. The problem is first formalized, followed by a detailed description of the recurrent neural network architecture, the batched training strategy for enhanced generalization, the formulation of the physics-informed loss function, and the optimization procedure.

Consider the FDDEs of the general form:

$${}_{a}^{C}\mathcal{D}_{x}^{\alpha}u(x) = f(x, u(x), u(x-\tau), u(\lambda x))), x \in (0, b],$$

subject to a history or initial condition defined over a preceding interval,

$$u(x) = g(x), x \in [a, 0],$$

where ${}_{a}^{C}\mathcal{D}_{x}^{\alpha}u(x)$ is the Caputo fractional derivative of order $0 < \alpha \leq 1, f$ is a prescribed function, $\tau > 0$ is a constant delay, $0 < \lambda < 1$ is a proportional delay factor, and $g(x)$ is a known history function. The objective is to develop a neural network that approximates the unknown solution $u(x)$ over the entire domain $[a, b]$.

To capture the inherent sequential and memory-dependent nature of FDDEs, a recurrent architecture is posited to be more suitable than a standard feed-forward network. While the recurrent nature of LSTMs is a natural fit for delay equations, a key question arises: is this sequential processing strictly necessary for achieving high accuracy, or does the LSTM's internal architecture itself provide an advantage? To investigate this, two distinct training strategies are explored: a fully recurrent approach and a non-recurrent approach that isolates the architectural benefits of the LSTM cell. It is hypothesized that the LSTM's internal gating mechanisms provide an intrinsic advantage over standard feed-forward networks, independent of their sequential application. A solution approximator,$\hat{u}(x; \boldsymbol{\omega})$, is proposed, consisting of an LSTM network followed by a fully connected layer, parameterized by the weights and biases $\boldsymbol{\omega} = \{\boldsymbol{W}_f, \boldsymbol{U}_f, \boldsymbol{b}_f, \boldsymbol{W}_i, \boldsymbol{U}_i, \boldsymbol{b}_i, \boldsymbol{W}_o, \boldsymbol{U}_o, \boldsymbol{b}_o, \boldsymbol{W}_c, \boldsymbol{U}_c, \boldsymbol{b}_c, \boldsymbol{W}_{fc}, \boldsymbol{b}_{fc}\}$. The core operations of an LSTM cell at each step j in a sequence are given by

$\mathbf{f}_j = \sigma(\mathbf{W}_f\mathbf{x}_j + \mathbf{U}_f\mathbf{h}_{j-1} + \mathbf{b}_f)$(Forget Gate)
$\mathbf{i}_j = \sigma(\mathbf{W}_i\mathbf{x}_j + \mathbf{U}_i\mathbf{h}_{j-1} + \mathbf{b}_i)$(Input Gate)
$\mathbf{o}_j = \sigma(\mathbf{W}_o\mathbf{x}_j + \mathbf{U}_o\mathbf{h}_{j-1} + \mathbf{b}_o)$(Output Gate)
$\tilde{\mathbf{c}}_j = \tanh(\mathbf{W}_c\mathbf{x}_j + \mathbf{U}_c\mathbf{h}_{j-1} + \mathbf{b}_c)$(Candidate Cell State)

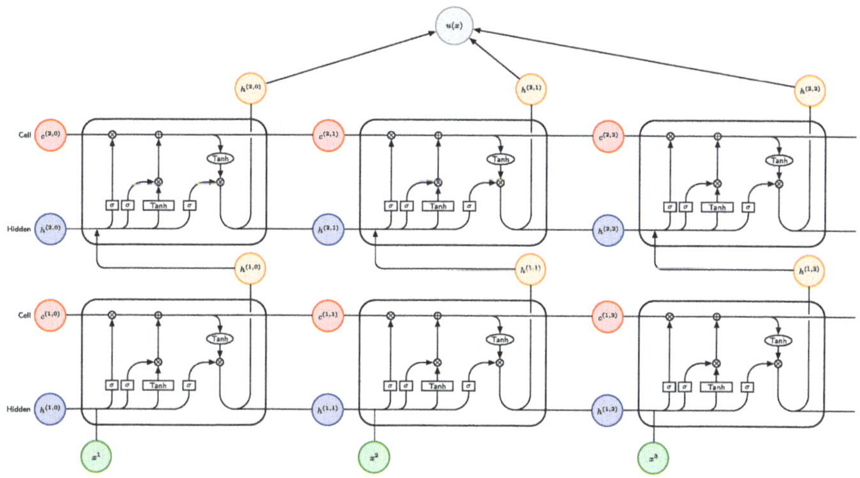

Fig. 1 The approximate solution of an FDDE using two stacked LSTMs with a sequence length of 3

$\mathbf{c}_j = \mathbf{f}_j \odot \mathbf{c}_{j-1} + \mathbf{i}_j \odot \tilde{\mathbf{c}}_j$(Cell State Update)

$\mathbf{h}_j = \mathbf{o}_j \odot \tanh\left(\mathbf{c}_j\right)$(Hidden State Update)where $\mathbf{x}_j$ is the input at step j, $\mathbf{h}_j$ and $\mathbf{c}_j$ are the hidden and cell states, respectively, and σ is the sigmoid function. The final scalar solution is obtained by passing the hidden state through an activation function and a linear transformation: $\hat{u}\left(x_j; \boldsymbol{\omega}\right) = \mathbf{W}_{fc}\tanh\left(\mathbf{h}_j\right) + \mathbf{b}_{fc}$. Figure 1 shows this process.

Two distinct training methodologies are investigated to evaluate the effectiveness of the LSTM architecture for this problem class. The first methodology, a point-wise approach, isolates the architectural benefit of the LSTM cell's internal gating mechanisms. For this purpose, a single problem domain is discretized into a set of $N + 1$ collocation points. Each point is treated as an independent sample within a large batch, forming an input tensor whose dimensions correspond to (batch size, sequence length, features) of $((N + 1) \times 1 \times 1)$. The propagation of the hidden state between consecutive domain points is intentionally circumvented by this configuration, forcing the network to operate without sequential memory.

The second methodology, a sequential approach, fully leverages the network's recurrent nature to learn generalized dynamics. A set of $N_{samples}$ distinct upper bounds $\left\{b_1, b_2, \ldots, b_{N_{samples}}\right\}$ is defined to create multiple problem instances. For each bound b_i, a corresponding sequence of points $[x_0, x_1, \ldots, x_N]^T$ is generated. The individual sequences are then stacked along the batch dimension to form a composite input tensor of shape $\left(\left(N_{samples}\right) \times (N + 1) \times 1\right)$. Training on this composite batch compels the model to learn the underlying dynamics of the FDDE, independent of a specific integration domain.

The network parameters $\boldsymbol{\omega}$ for both training methodologies are optimized by minimizing a composite loss function that enforces both the history condition and the governing differential equation. The total loss $\mathcal{L}(\boldsymbol{\omega})$ is a weighted sum of two components:

$$\mathcal{L}(\omega) = w_{hist}\mathcal{L}_{hist}(\omega) + w_{res}\mathcal{L}_{res}(\omega),$$

where w_{hist} and w_{res} are user-defined weights.

The history loss, $\mathcal{L}_{hist}$, measures the discrepancy between the network prediction and the known history function $g(x)$ on the interval $[a, 0]$. It is defined as the mean squared error over all collocation points $\mathbf{x}_{hist}$ in the history domain for all samples in the batch:

$$\mathcal{L}_{hist}(\omega) = \frac{1}{N_{samples}} \sum_{i=1}^{N_{samples}} ; \frac{1}{|\mathbf{x}_{hist,i}|} \sum_{x_j \in \mathbf{x}_{hist,i}} \left(\hat{u}\left(x_j; \omega\right) - g\left(x_j\right)\right)^2.$$

The residual loss, $\mathcal{L}_{res}$, enforces the FDDE itself. For each sample i in the batch, the residual is defined at collocation points $\mathbf{x}_{ode,i}$ in the domain $(0, b_i]$ as follows:

$$\mathcal{R}\left(x_j; \omega\right) = \left(\mathcal{M}_i \hat{\mathbf{u}}_{ode,i}\right)_j - f\left(x_j, \hat{u}\left(x_j; \omega\right), \hat{u}\left(x_j - \tau; \omega\right), \hat{u}\left(\lambda x_j; \omega\right)\right),$$

where $\hat{\mathbf{u}}_{ode,i}$ is the vector of network predictions on the grid $\mathbf{x}_{ode,i}$, and $\mathcal{M}_i$ is the precomputed Caputo operational matrix corresponding to that specific grid.

The evaluation of delayed terms, such as $\hat{u}\left(x_j - \tau\right)$ or $\hat{u}\left(\lambda x_j\right)$, is a critical component of the residual calculation. For constant delays where τ is an integer multiple of the grid spacing, the argument $x_j - \tau$ aligns with a previous grid point x_{j-k}. th This term is evaluated efficiently using direct indexing of the network's output vector:$\hat{u}\left(x_j - \tau\right) = \hat{u}\left(x_{j-k}\right)$. Proportional delays, such as λx_j, typically produce arguments that fall between collocation points. Such cases can be handled using two strategies. The first strategy re-evaluates the network with a new forward pass at the off-grid point λx_j. This approach is highly accurate but introduces significant computational overhead. A second, more efficient strategy uses linear interpolation between the network's existing outputs at the two nearest grid points, x_k and x_{k+1}:

$$\hat{u}\left(\lambda x_j\right) \approx \frac{x_{k+1} - \lambda x_j}{x_{k+1} - x_k}\hat{u}(x_k) + \frac{\lambda x_j - x_k}{x_{k+1} - x_k}\hat{u}(x_{k+1}).$$

The residual loss of the FDDE is the mean squared error of these residuals, averaged over all samples:

$$\mathcal{L}_{res}(\omega) = \frac{1}{N_{samples}} \sum_{i=1}^{N_{samples}} \frac{1}{|\mathbf{x}_{ode,i}|} \sum_{x_j \in \mathbf{x}_{ode,i}} \mathcal{R}(x_j; \omega)^2.$$

The minimization of the loss function $\mathcal{L}(\omega)$ is performed using gradient-based optimization. The gradients of the loss with respect to the network parameters, $\nabla_\omega \mathcal{L}$, are computed efficiently via backpropagation. While first-order methods such as

Adam are commonly employed, quasi-Newton methods like the Broyden–Fletcher–Goldfarb–Shanno (L-BFGS) algorithm often yield faster convergence for scientific machine learning problems. The L-BFGS optimizer approximates the inverse Hessian matrix to inform its update rule, which at iteration k is

$$\omega_{k+1} = \omega_k - \eta_k \mathbf{H}_k \nabla_\omega \mathcal{L}(\omega_k),$$

where η_k is a step size determined by a line search (e.g., strong Wolfe conditions) and $\mathbf{H}_k$ is the approximate inverse Hessian. The next section provides some numerical examples using the proposed method.

4 Numerical Results

Numerical results are presented to validate the proposed physics-informed LSTM framework. The framework is implemented in PyTorch, with the *pinnies* [24] library employed to construct the Caputo operational matrix. All computations were performed on a Google Colab instance with a NVIDIA T4 GPU. A preliminary hyper-parameter search indicated that an LSTM architecture with a low number of hidden layers and a hidden size of less than 100 consistently yielded accurate results; the model's performance exhibited low sensitivity to minor variations within this range. The proposed methods are applied to several representative FDDE problems in the following examples.

Example 1 Consider the fractional delay differential equation.

$$\mathcal{D}_x^{0.5} u(x) = u(x-1) + f(x), x \in (0, 2],$$

where $\mathcal{D}_x^{0.5}$ denotes the Caputo fractional derivative of order 0.5 and the delay is $\tau = 1$. The forcing term is

$$f(x) = -x \frac{-3\sqrt{\pi}x + 8\sqrt{x} + 6\sqrt{\pi}}{3\sqrt{\pi}}.$$

The history function on $[-1, 0]$ is

$$g(x) = 1 - x^2, x \leq 0.$$

The forcing term $f(x)$ and history function aer selected so that the exact analytical solution is $u_{\text{exact}}(x) = 1 - x^2$ over the entire domain. This problem is solved using both training approaches outlined in Sect. 3.

First, the training approach that circumvents sequential processing is employed. The domain $[-1, 2]$ is discretized into $N = 1000$ points and the input is structured as a single batch in which each point serves as an independent sample. The model, an LSTM with 3 layers and a hidden size of 15, is trained for 150 iterations using the L-BFGS optimizer. Figure 2 compares the predicted and exact solutions, which show excellent agreement. The pointwise absolute error, $|u_{\text{exact}}(x) - \hat{u}(x)|$, in Fig. 3 confirms the high accuracy of this method for a fixed domain.

Next, the recurrent, batched generalization method is evaluated. A batch of $N_{samples} = 10$ problem instances is constructed, where each instance shares the same history domain $[-1, 0]$ but has a different upper bound b_i, linearly spaced from 0.2 to 2.0. The network processed these 10 trajectories in parallel, learning a single set of weights to solve the entire family of problems. Figure 4 plots the predicted solution curves for several of these domains against the exact solution. The model accurately captures the solution over all trained domains, which confirms its ability

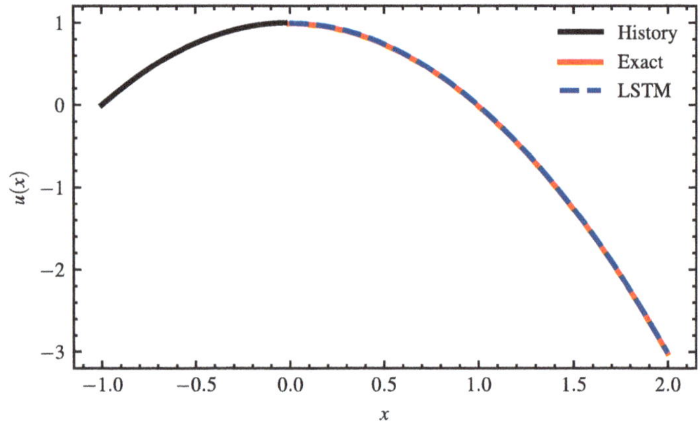

Fig. 2 Predicted vs. exact solution for Example 1

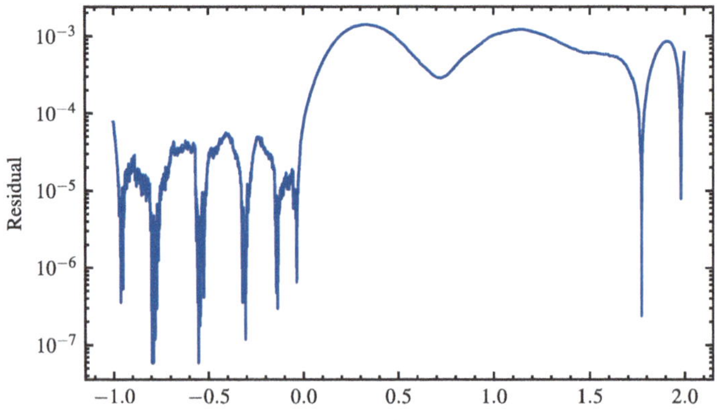

Fig. 3 Pointwise absolute error (log-scale) for Example 1

to learn the underlying dynamics. Figure 5 shows the absolute error for the longest trajectory ($b = 2.0$), which remains low across the domain and is comparable to the error from the first method.

The extrapolation capabilities of the two training approaches are further compared. Both models are trained on the domain $[-1, 2]$ and their predictions are evaluated on an unseen extended domain up to $x = 2.5$. Figure 6 and Fig. 7 show that both models provide reasonable extrapolations beyond the training boundary. The batched generalization method, however, requires significantly more computational resources because it processes multiple data sequences, and also due to the sequential nature of LSTMs. The first approach achieves comparable performance at a lower computational cost; this method is therefore adopted for the remaining examples.

Example 2 Let the nonlinear fractional delay differential equation.

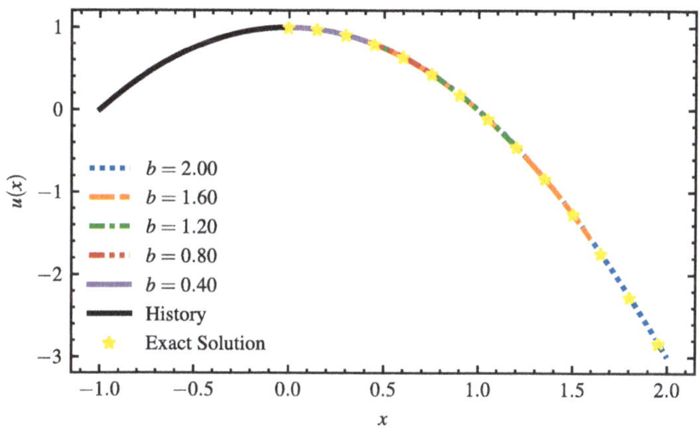

Fig. 4 Predicted solutions for multiple domains for Example 1

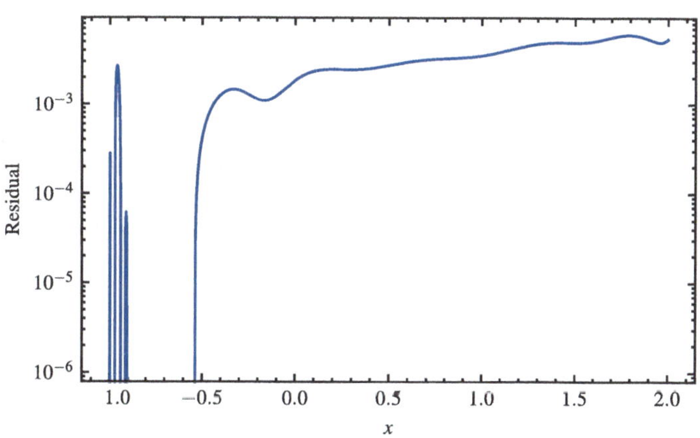

Fig. 5 Absolute error for the b = 2.0 case (log-scale) for Example 1

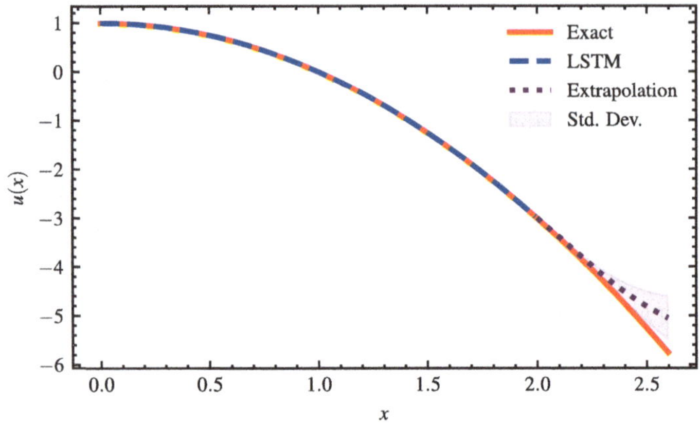

Fig. 6 Extrapolation from the non-recurrent point-wise model for example 1

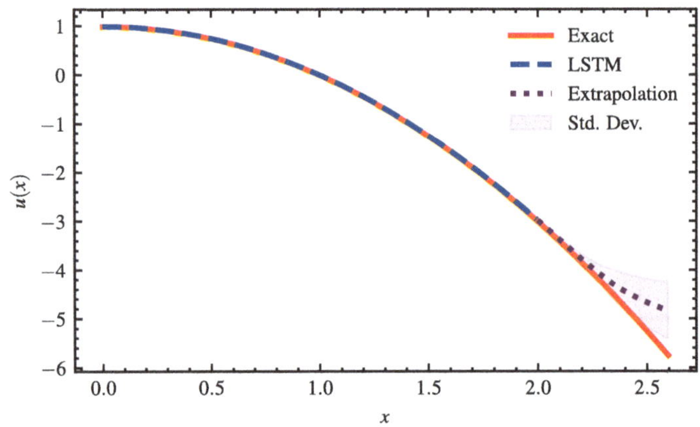

Fig. 7 Extrapolation from the recurrent sequential model for example 1

$$\mathcal{D}_x^{0.5}u(x) - u\left(\frac{x}{3}\right) - u(x)^2 = f(x), x \in [0, 2],$$

where $\mathcal{D}_x^{0.5}$ is the Caputo fractional derivative of order 0.5. The proportional delay argument $x/3$ rarely coincides with the discrete training grid. Although such values can be predicted by performing another forward pass of the network, $u(x/3)$ is evaluated using an interpolation of the network's outputs at nearby grid points. The forcing term is

$$f(x) = \frac{2(-3+4x)\sqrt{x}}{3\sqrt{\pi}} - \frac{x^2}{9} + \frac{x}{3} - (x^2 - x)^2,$$

with the initial condition $u(0) = 0$. The exact solution is $u_{\text{exact}}(x) = x^2 - x$.

Physics-informed LSTM network with three layers and 32 hidden units is used. The model is trained for 150 iterations using the L-BFGS optimizer on a grid of $N = 1000$ points over $[0, 2]$. The loss function combines the initial condition and the equation residual, with the Caputo derivative approximated using the proposed fractional differentiation matrix. Figure 8 compares the predicted and exact solutions, which show excellent agreement. Figure 9 displays the pointwise absolute error on a logarithmic scale. The results show the LSTM-PINN accurately resolves the nonlinear fractional delay dynamics.

A convergence comparison between the non-recurrent LSTM approach and a standard multi-layer perceptron (MLP) directly tests the hypothesis that the LSTM's internal architecture provides an advantage even without sequential processing. An MLP with a $[1, 70, 70, 1]$ architecture, having a comparable number of trainable

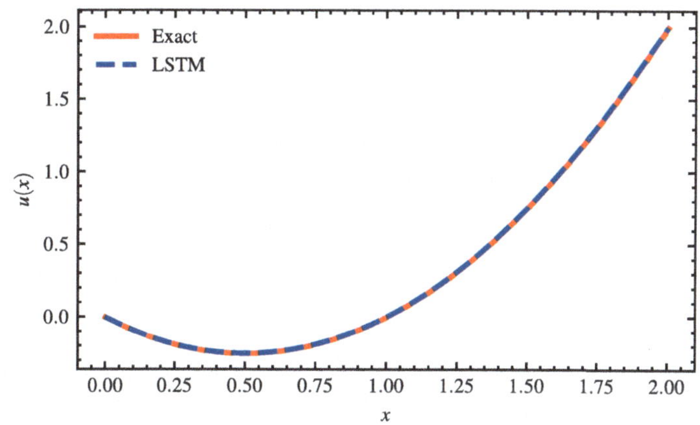

Fig. 8 Predicted vs. exact solution for Example 2

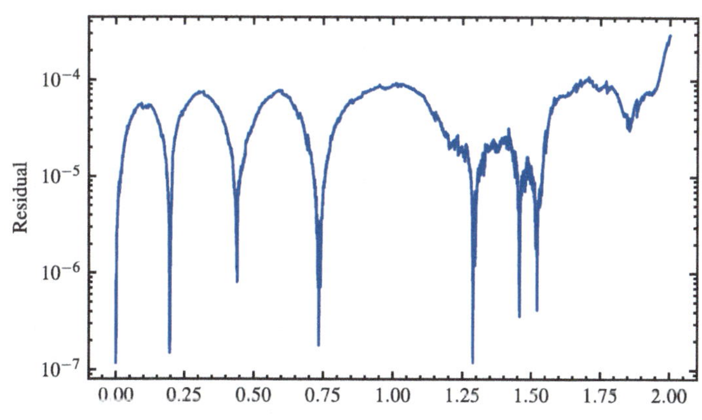

Fig. 9 Pointwise absolute error (log-scale) for Example 2

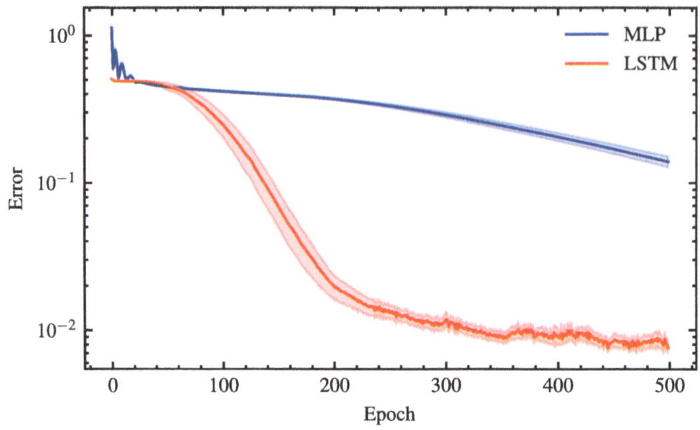

Fig. 10 Convergence history of the mean absolute error for the MLP and LSTM models, averaged over 100 runs. The shaded regions represent the 95% confidence interval

parameters to the LSTM model, is employed, following the solution method from Taheri et al. [13]. Both networks are trained with the Adam optimizer, and the mean absolute error is recorded over 100 independent runs. Figure 10 shows that the non-recurrent LSTM converges significantly faster and to a lower error than the MLP.

The superior performance of the non-recurrent LSTM can be attributed to its internal architectural design. Unlike a standard MLP, where gradients must propagate backward through all layers uniformly, the LSTM's gating mechanisms regulate information flow. The input and forget gates create adaptive pathways for gradients, which can mitigate the vanishing and exploding gradient problems that often hinder the training of deep feed-forward networks. This leads to a more stable and efficient optimization process. Furthermore, the LSTM's cell state acts as an internal memory channel. Even in a non-recurrent setup, this allows the network to learn more complex and non-local relationships within the solution space. While an MLP approximates the solution globally, the LSTM can be viewed as learning a more adaptive representation at each point, which makes it better suited for capturing the intricate dynamics of fractional differential equations. The observed faster convergence is therefore not merely an empirical result but a consequence of an architecture inherently better equipped for this class of problems.

Example 3 Consider the fractional delay differential equation:

$$\mathcal{D}_x^{0.3}u(x) = u(x-1) - u(x) + \left(1 - 3x + 3x^2 + \frac{2000x^{2.7}}{1071\Gamma(0.7)}\right), x \in (0, 1],$$

where $\mathcal{D}_x^{0.3}$ is the Caputo derivative of order 0.3. The history function on $[-1, 0]$ is $g(x) = x^3$. The exact analytical solution is $u_{\text{exact}}(x) = x^3$ for $x \geq 0$.

An LSTM network with three hidden layers and 15 hidden units is employed. The model was trained for 150 iterations with the Adam optimizer on a grid of $N =$

1000 points over the domain $[-1, 1]$. Figure 11 compares the predicted and exact solutions, and Fig. 12 presents the pointwise absolute error on a logarithmic scale. The results confirm that the LSTM-based method accurately captures the fractional delay dynamics. To further assess performance, the method is validated against the Haar wavelet collocation technique (HWCT) [14] and the fractional backward difference method for delay equations [15]. The results in Table 1 show that the proposed LSTM-based approach achieves superior accuracy.

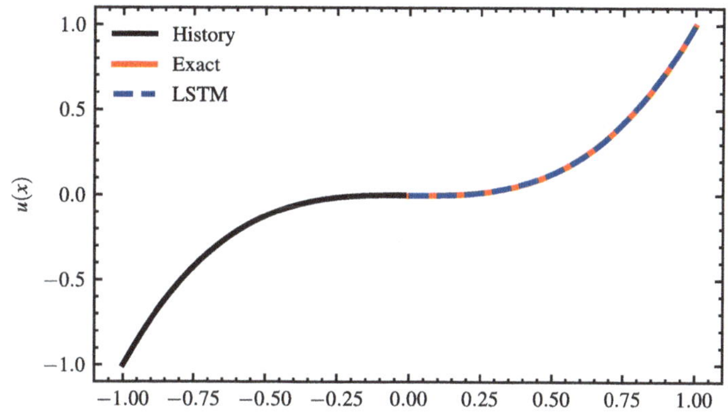

Fig. 11 Predicted vs. exact solution for Example 3

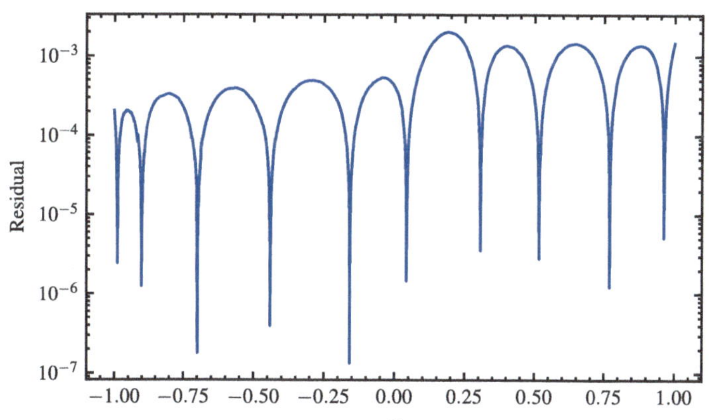

Fig. 12 Pointwise absolute error (log-scale) for Example 3

Table. 1 Error comparison for Example 3

	HWCT [14]	Method [15]	Proposed LSTM-based method
L_∞ error	2.81×10^{-3}	7.10×10^{-2}	7.45×10^{-4}

Example 4 Consider a fractional delay differential equation with a piecewise analytical solution:

$$\mathcal{D}_x^{0.5} u(x) = u(x-1) + f(x), x \in (0, 2],$$

with the history condition $u(x) = 1$ for $x \leq 0$. The forcing term $f(x)$ is a piecewise function,

$$f(x) = \frac{1}{3\sqrt{\pi}} \times \begin{cases} 8x^{3/2} - 3\sqrt{\pi} & x \leq 1 \\ 8x^{3/2} + \frac{16(x-1)^{5/2}}{5} + \frac{(-15x^2+30x-30)\sqrt{\pi}}{5} & 1 < x \leq 2 \\ 8x^{3/2} + \frac{16(x-1)^{5/2}}{5} + \frac{(-5x^3+15x^2-30x+10)\sqrt{\pi}}{5} & x > 2 \end{cases}$$

The forcing term is chosen so that the exact solution is the piecewise function:

$$u_{\text{exact}}(x) = \begin{cases} 1, & x \leq 0, \\ x^2 + 1, & 0 < x \leq 1, \\ \frac{x^3}{3} + x + \frac{2}{3}, & x > 1. \end{cases}$$

An LSTM network with three layers and 64 hidden units is employed. The model is trained for 150 epochs using the L-BFGS optimizer on a grid of $N = 1000$ points over the domain $[-1, 2]$. The loss function combines the equation residual and the history condition.

Figure 13 compares the predicted and exact solutions. The LSTM prediction aligns closely with the exact piecewise profile and captures the derivative discontinuity at $x = 1$. Figure 14 displays the pointwise absolute error on a logarithmic scale, which confirms the model's accuracy across the different segments of the solution.

Example 5 A fractional delay differential equation is given by.

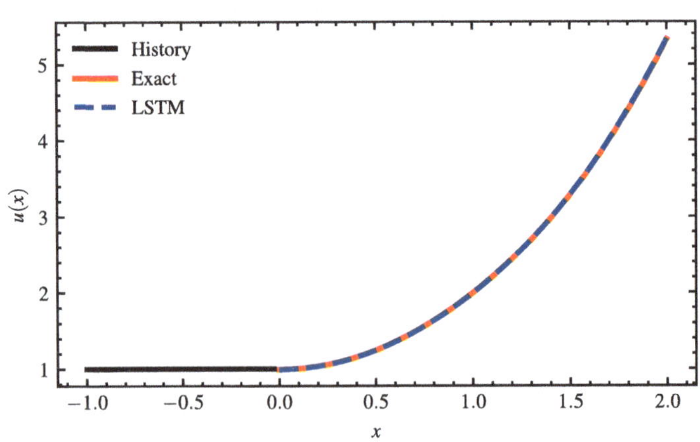

Fig. 13 Predicted vs. exact solution for Example 4

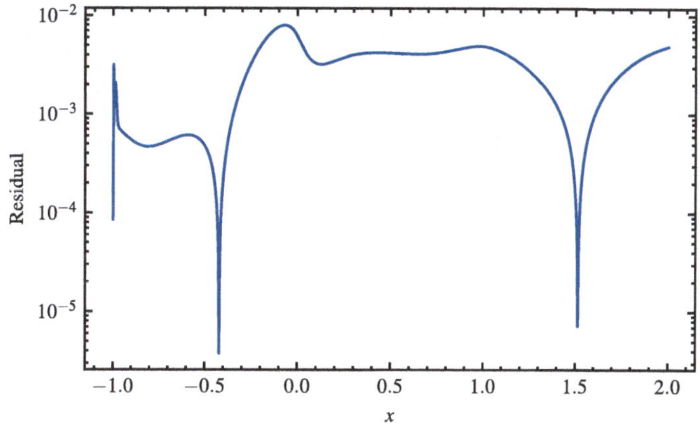

Fig. 14 Pointwise absolute error (log-scale) for Example 4

$$\mathcal{D}_x^{0.5}u(x) = u(x-1) + f(x),$$

on the domain $x \in (0, 3]$ with a delay $\tau = 1$. The exact solution is $u_{\text{exact}}(x) = e^{-\lambda x}\sin(\omega x)$ with parameters $\omega = 10$ and $\lambda = 1.75$. The forcing term $f(x)$ is constructed so that the analytical solution satisfies the equation exactly. The history condition for $x \leq 0$ is given by the exact solution.

An LSTM network with three layers and 15 hidden units is employed. The domain $[-1, 3]$ is discretized with $N = 2000$ points and a Caputo derivative matrix of order 0.5 is used. The network is trained in two stages: first for 4000 iterations with the Adam optimizer, followed by refinement with the L-BFGS optimizer for 200 epochs.

Figure 15 compares the predicted and exact solutions, which show excellent agreement even for this highly oscillatory case. Figure 16 presents the pointwise absolute error on a logarithmic scale which confirms the model's accuracy.

5 Conclusions

A physics-informed framework is developed to solve fractional delay differential equations. The method employs a LSTM network as the solution approximator and integrates a precomputed operational matrix for the Caputo fractional derivative. Such a design combines the architectural strength of recurrent networks for modeling memory-dependent systems with a computationally efficient method for handling non-local fractional operators.

The framework is validated on several benchmark problems exhibiting nonlinear, piecewise, and highly oscillatory dynamics. The experiments reveal that a non-recurrent LSTM training configuration achieves accuracy comparable to a fully recurrent approach while offering lower computational cost and greater flexibility.

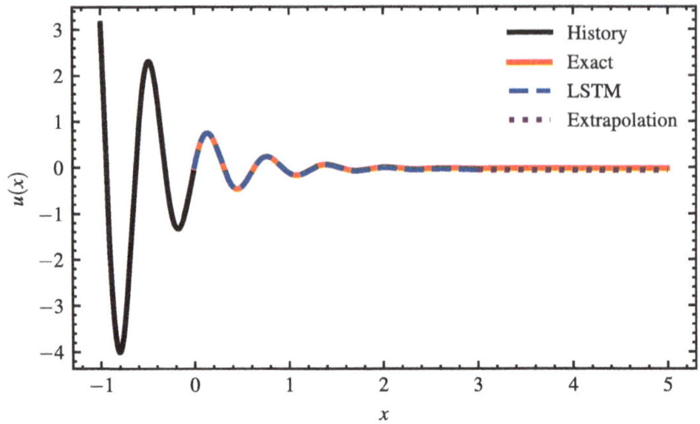

Fig. 15 Predicted vs. exact solution for Example 5

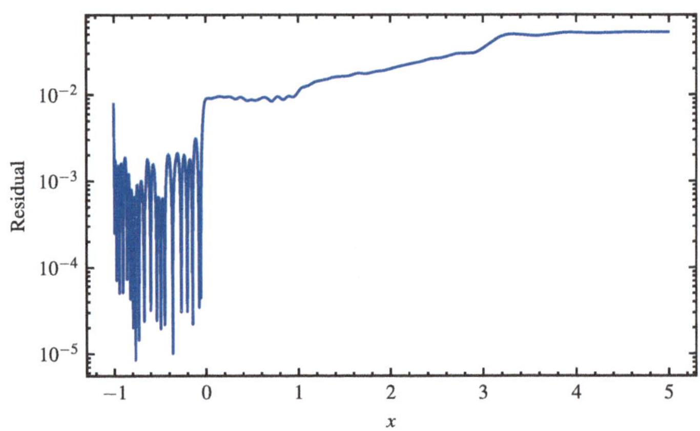

Fig. 16 Pointwise absolute error (log-scale) for Example 5

The superior performance of this non-recurrent configuration over a standard MLP suggests that the architectural benefits of the LSTM for this class of problems stem not only from its ability to process sequences but also from the rich optimization landscape provided by its internal gating mechanisms. The results also confirm that the precomputed operational matrix provides a computational advantage over direct numerical integration.

Declarations **Conflict of Interests:** The authors declare that there are no conflicts of interest.

Funding None.

Data Availability The authors confirm that the data supporting the findings of this study are available within the article.

CRediT Authorship **Tayyebeh Taheri:** Writing—Original Draft, Formal Analysis, Writing—Review & Editing, Software, Resources, Investigation, Methodology;

Pegah Ahadian: Writing—Original Draft, Resources;
Ali Mehrban: Writing—Original Draft, Software;
Alireza Afzal Aghaei: Methodology, Software, Conceptualization, Writing—Review & Editing, Supervision.

References

1. Balachandran, B., Kalmár-Nagy, T., Gilsinn, D.E.: Delay Differential Equations. Springer, Berlin (2009)
2. Derfel, G.A.: Kato problem for functional-differential equations and difference Schrödinger operators. Oper. Theory **46**(2), 319–321 (1990)
3. Boubaker, K., Van Gorder, R.A.: Application of the BPES to Lane-Emden equations governing polytropic and isothermal gas spheres. New Astron. **17**(6), 565–569 (2012)
4. Erneux, T.: Applied Delay Differential Equations. Springer, New York (2009)
5. Owoyele, O., Pal, P.: ChemNODE: A neural ordinary differential equations framework for efficient chemical kinetic solvers. Energy AI **7**, 100118 (2022)
6. Chishti, F., Hanif, F., Shams, R.: A comparative study on solution methods for fractional order delay differential equations and its applications. Math. Sci. Appl. **2**(1), 1–13 (2023)
7. Yi, S., Ulsoy, A.G., Nelson, P.W.: Solution of systems of linear delay differential equations via Laplace transformation. In: *Proc. 45th IEEE Conf. on Decision and Control*, IEEE, pp. 1–6 (2006)
8. Johnson, A.: Investigation of network models finite difference method. Eurasian J. Chem. Med. Pet. Res. **2**(1), 1–9 (2023)
9. Khasawneh, F.A., Barton, D.A.W., Mann, B.P.: Periodic solutions of non-linear delay differential equations using spectral element method. Nonlinear Dyn. **67**(1), 641–658 (2012)
10. Raissi, M., Perdikaris, P., Karniadakis, G.E.: Physics-informed neural networks: A deep learning framework for solving forward and inverse problems involving nonlinear partial differential equations. J. Comput. Phys. **378**, 686–707 (2019)
11. Hochreiter, S., Schmidhuber, J.: Long short-term memory. Neural Comput. **9**(8), 1735–1780 (1997)
12. Hagge, T., et al.: Solving differential equations with unknown constitutive relations as recurrent neural networks. *arXiv preprint* arXiv:1710.02242 (2017)
13. Taheri, T., Afzal Aghaei, A., Parand, K.: Accelerating fractional PINNs using operational matrices of derivative. *arXiv preprint* arXiv:2401.14081 (2024)
14. Amin, R., et al.: A computational algorithm for the numerical solution of fractional order delay differential equations. Appl. Math. Comput. **402**, 125863 (2021)
15. Morgado, M.L., Ford, N.J., Lima, P.M.: Analysis and numerical methods for fractional differential equations with delay. J. Comput. Appl. Math. **252**, 159–168 (2013)
16. Taheri, T., Afzal Aghaei, A., Parand, K.: Bridging machine learning and weighted residual methods for delay differential equations of fractional order. Appl. Soft Comput. **149**, 110936 (2023)
17. Taheri, T., Afzal Aghaei, A., Parand, K.: A new kernel-based approach for solving general fractional (integro)-differential-algebraic equations. Eng. Comput. **41**(2), 845–864 (2025)
18. Hopfield, J.J.: Hopfield network. Scholarpedia **2**(5), 1977 (2007)
19. Nonnenmacher, T.F., Metzler, R.: On the Riemann-Liouville fractional calculus and some recent applications. Fractals **3**(03), 557–566 (1995)
20. Oliveira, D.S., de Oliveira, E.C.: On a Caputo-type fractional derivative. Adv. Pure Appl. Math. **10**(2), 81–91 (2019)
21. Sivalingam, S.M., Kumar, P., Govindaraj, V.: A neural networks-based numerical method for the generalized Caputo-type fractional differential equations. Math. Comput. Simul **213**, 302–323 (2023)

22. Sivalingam, S.M., Kumar, P., Govindaraj, V.: A novel optimization-based physics-informed neural network scheme for solving fractional differential equations. Eng. Comput. **40**(2), 855–865 (2024)
23. Sivalingam, S.M., Kumar, P., Govindaraj, V.: A Chebyshev neural network-based numerical scheme to solve distributed-order fractional differential equations. Comput. Math. Appl. **164**, 150–165 (2024)
24. Aghaei, A.A., Movahedian Moghaddam, M., Parand, K.: Pinnies: an efficient physics-informed neural network framework to integral operator problems. *arXiv preprint* arXiv:2409.01899 (2024)

Approximate Numerical Solutions of Second-Order Delay Differential Equations Using Artificial Neural Networks

Khadeejah James Audu, Falade Kazeem Iyanda, Oloruntobi Olalekan Balogun, Abd'Gafar Tunde Tiamiyu, Olufemi Adesola Adedayo, Ayodele Victoria Iyadunni, and Adio Adesina Kamorudeen

Abstract The study presents an innovative and efficient solution method for Second-Order Delay Differential Equations (SODDEs) through the use of Artificial Neural Networks (ANNs), also known as Multilayer Perceptrons (MLP). The research introduces a fully ANN-based method to overcome accuracy problems in time-delayed systems that affect traditional numerical methods such as Runge–Kutta and Euler. The core innovation of this approach lies in embedding the history function directly into the ANN structure, thereby eliminating the need for hybrid numerical methods. The proposed ANN method demonstrates exceptional accuracy in solving SODDE problems through detailed testing against exact solutions, the fourth-order Runge–

K. J. Audu · O. O. Balogun
Department of Mathematics, Federal University of Technology, Minna, Nigeria
e-mail: k.james@futminna.edu.ng

O. O. Balogun
e-mail: oloruntobi.m1803525@st.futminna.edu.ng

F. K. Iyanda (✉)
Department of Mathematics, Aliko Dangote University of Science and Technology, Wudil Kano State, Nigeria
e-mail: faladekazeem2016@kustwudil.edu.ng

A. T. Tiamiyu
Department of Mathematics, Chinese University of Hong Kong, Sha Tin, Hong Kong

O. A. Adedayo
Department of Mathematics and Statistics, University of Massachusetts, Amherst, Massachusetts (MA), USA
e-mail: oadedayo@umass.edu

A. V. Iyadunni
Department of Mathematics, University of Dundee, Dundee, Scotland, UK

A. A. Kamorudeen
Department of Basic Sciences, School of Science and Technology, Babcock University, Ilishan-Remo, Ogun State, Nigeria
e-mail: adioa@babcock.edu.ng

L. Kaur and P. Kumar (eds.), *Neural Networks: Theory, Algorithms, Simulation, and Applications*, Studies in Computational Intelligence 1273,
https://doi.org/10.1007/978-3-032-18750-5_6

Kutta, and Euler methods, resulting in near-zero error. The optimization and error analyses confirm both robust convergence and efficient computation. The ANN-based approach shows strong potential to become an accurate and reliable solution method for SODDEs, enabling predictive modeling and dynamic system analysis across disciplines with critical delayed dynamics.

Keywords Artificial neural networks · Second-order delay differential equations · Universal approximation · Computational efficiency · Predictive modeling

1 Introduction

Differential equations are fundamental to modeling dynamic systems across scientific disciplines, providing essential tools for understanding and predicting complex phenomena in fields such as physics, biology, engineering, and economics [1]. These equations describe how quantities change with time or space, allowing researchers to model various processes in areas such as mineral processing machines [2], population dynamics [3], financial markets [4], and electrical circuits [5]. Due to their ability to model multiple changing variables, differential equations enable researchers to enhance processes, develop innovative solutions, and gain deeper insights into the dynamic systems that form the foundation of technological, medical, and environmental advancements [6].

Second-Order Delay Differential Equations (SODDEs) belong to this general classification but are particularly complex and more challenging than many other types. These equations include second derivatives and time-delay terms, which capture how a system evolves based not only on its current state but also on past states [7]. This feature makes SODDEs especially effective for modeling systems whose present behavior depends heavily on historical data, such as control systems [8], population dynamics [9], and pharmacokinetics [10]. However, time delays inevitably complicate the computational aspects of SODDEs, making their solutions difficult to obtain [11]. Various conventional numerical techniques, such as the Runge–Kutta method [12], block-step methods [13], and the Adomian Decomposition Method (ADM) [14], have been used to approximate the solutions of SODDEs. Although these approaches can be applied to a wide range of problems, they often suffer from instability and inaccuracies, particularly in complex or highly delayed systems [15].

Moreover, these numerical methods can be computationally expensive, especially when applied to high-dimensional systems, as noted in [16]. To overcome these limitations, there has been growing interest in developing Artificial Neural Networks (ANNs) because they offer a flexible and efficient means of approximating solutions to differential equations [17]. Given their capacity to learn mathematical function approximations, ANNs have been employed to solve Ordinary Differential Equations (ODEs) [18–21] and Partial Differential Equations (PDEs) [22]. Owing to their universal approximation property [23], ANNs can approximate the functional dependencies inherent in differential equations; in some cases, they have demonstrated better prediction and simulation capabilities than traditional numerical methods [24].

Consequently, the integration of ANNs with other numerical methods, such as the finite element method [25] and the finite difference method [26], has shown promising results in terms of computational efficiency and accuracy [27]. Nevertheless, there remains a noticeable lack of publications focusing on the application of a purely ANN-based approach to solving second-order DDEs [28]. Most previous studies have relied on hybrid methods in which ANNs are combined with conventional numerical techniques, thereby limiting the full realization of the ANN's versatility [29]. Although these hybrid approaches are effective, they only partially exploit the potential of ANNs for handling delayed dynamics in isolation [30]. While numerous studies have employed methods such as block steps [31], ADM [32], or Runge–Kutta schemes [33] to numerically solve Second-Order Delay Differential Equations (SODDEs), no existing work has utilized a Multilayer Perceptron (MLP) neural network approach that relies solely on the history function as the basis for estimating second-order DDE solutions. Although the authors in [38] applied Physics-Informed Neural Networks (PINNs) to first-order delay differential equations, this paper introduces a neural network approach for solving second-order delay differential equations using ANNs, which possess universal approximation capability. By directly incorporating the history function, this work aims to overcome the computational challenges associated with SODDEs and present a novel approach to solving these equations. This innovative method may contribute to advancing the use of SODDEs in predictive modeling and other scientific applications, providing a foundation for more accurate simulations in systems with delayed dynamics.

The significance of this study lies in its contribution to the body of literature on computational techniques for solving SODDEs using ANNs. In doing so, it removes barriers associated with traditional numerical methods that limit the ability to analyze dynamic systems characterized by time delays and second-order derivatives. Since such systems are prevalent in control theory, population dynamics, and pharmacokinetics, the development of a purely ANN-based approach has the potential to revolutionize computational methods across various disciplines. Furthermore, this research enriches current studies on the application of ANNs to differential equations and demonstrates their capacity to independently address the complexities of delayed dynamics, thereby advancing the field of predictive modeling and real-time simulation.

2 Theoretical Foundation

Definition 1 *Artificial Neural Network:* Artificial Neural Network: An artificial neural network is a computational architecture that consists of multiple layers: an input layer, one or more hidden layers, and an output layer. Let be the total number of layers and let the number of neurons in each of those layers be $(M_0, M_1, \cdots, M_H)$.

M_0 represents the dimensionality of the inputs, M_H defines the dimensionality of the output, and M_h signifies the number of neurons in the $h-th$ hidden layer for $h \in [1, H-1]$.

The network's behavior is determined by a set of parameters denoted as θ, which comprises weight matrices $W^{(h)}$ and bias vectors $b^{(h)}$ for each layer h. For each layer, the parameters are structured as follows:

$$\theta = \{\left(W^{(h)}, b^{(h)}\right) | h = 1, 2, \cdots, H\} \tag{1}$$

When an input vector $x \in \mathbf{R}^{M_0}$ is given to the network, the following actions take place. The first layer computes the input to the neurons as:

$$z^{(1)} = W^{(1)} \cdot x + b^1 \tag{2}$$

where $W^{(1)}$ is a weight matrix with dimensions of $M_1 \times M_0$ and $b^{(1)}$ is a bias of size M_1. Subsequent hidden layers apply a non-linear activation function $\sigma : \mathbf{R} \to \mathbf{R}$ to the linear combinations, producing:

$$a^{(h)} = \sigma(z^{(h)}), \qquad for \quad h \in [1, H-1] \tag{3}$$

where $z^{(h)}$ for the $h-th$ layer is given by:

$$z^{(h)} = W^{(h)} \cdot a^{(h-1)} + b^{(h)} \tag{4}$$

Finally, in the output layer, the computation is performed similarly, resulting in:

$$z^{(H)} = W^{(H)} \cdot a^{(H-1)} + b^{(H)} \tag{5}$$

and the final output from the network is:

$$a^{(H)} = \sigma(z^{(H)}) \tag{6}$$

Definition 2 A second-order delay differential equation is defined as [35]

$$u^{''} + a_1 u^{'}(t) + a_0 u(t) = f(t, u(t-\tau_1), u(t-\tau_2)) \tag{7}$$

with initial conditions:

$$u(t) = g_0(t), \qquad t \in [-\tau_2, 0] \tag{8}$$

$$u^{'}(t) = g_1(t), \qquad t \in [-\tau_1, 0] \tag{9}$$

where a_0, a_1, τ_1 and τ_2 are constants and $g_0(t)$ and $g_1(t)$ are continuous functions.

Lemma 1 *Existence of solution: Let $h \in C(J, \mathbf{R})$ then the solution to the problem is governed by the expression [33, 34]*

$$u'' = h(v) + C_1 u(v) + C_2 u(v - \alpha), \qquad v \in J \tag{10}$$

$$u(v) = \Psi(v), \quad -\alpha \leq v \leq 0, \quad u(0) = \beta_1, \quad u'(0) = \beta_2, v \in J \tag{11}$$

can be expressed as

$$u(v) = \beta_1 + v\beta_2 + \int_0^v (v - s)[h(s) + C_1 u(s) + C_2 u(s - \alpha)]ds, \qquad v \in J \tag{12}$$

Lemma 2 *Any bounded, measurable sigmoidal function is discriminatory. Specifically, any continuous sigmoidal function, including the sigmoid function $\sigma(x) = \frac{1}{1+e^{-x}}$ and the **ReLU** function **ReLU(x)** $= max(0, x)$, is also discriminatory.*

Theorem 1 *Universal Approximation Theorem (Cybenko) [31]: Let any continuous discriminatory function be denoted as σ. It follows that sums of the form*

$$Z(x) = \sum_{j=1}^{N} \alpha_j \sigma(w_j^T x + b_j) \tag{13}$$

can be constructed, where $w_j \in \mathbf{R}^n$ and both α_j and b_j belong to $\mathbf{R}$. These sums are dense in the space $C(I^n)$ i.e. that for any given $\hat{O} > 0$ and for any function $f \in C(I^n)$, it is possible to find a sum $Z(x)$ of this nature such that $|Z(x) - f(x)| < \hat{O}$ for all $x \in I^n$.

3 Methodology

3.1 Description of the Proposed ANN Technique

The artificial neural network used in this study consists of three hidden layers, each containing ten neurons. The neurons compute weighted sums of their inputs and apply nonlinear activation functions, namely the sigmoid and ReLU functions. Figure 1 illustrates the architecture of the proposed network, showing the input layer, hidden layers, and output layer.

Two types of activation functions were employed: the sigmoid function, which maps inputs to the range (0, 1), and the ReLU (Rectified Linear Unit) function, which out-puts the input directly if it is positive and zero otherwise. These activation functions were selected for their distinct advantages in capturing nonlinearity and enhancing learning dynamics during training. The network's error was evaluated using the mean squared error (MSE) loss function, which measures the difference

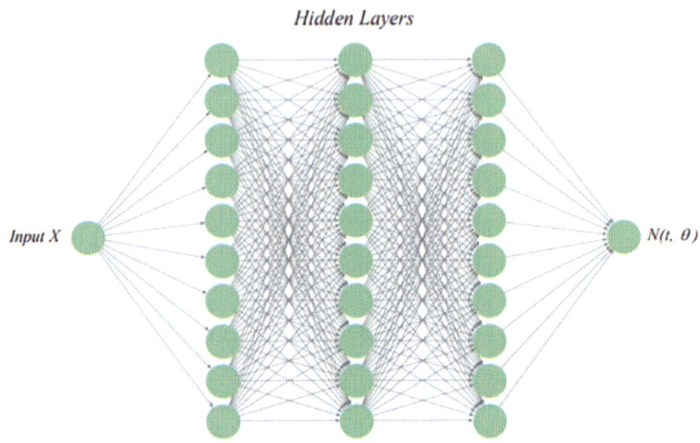

Fig. 1 Description of the Proposed ANN architecture

between the predicted and actual values and guides the adjustment of weights and biases to minimize prediction error throughout the training process.

3.2 Mathematical Formulation

The development of the ANN architecture begins by representing the solution as a neural network function that incorporates both a history function and an approximating function. In general, a second-order delay differential equation can be expressed in the form:

$$\ddot{u}(t) = f(t, u(t), \dot{u}(t)), u(t - \tau_1), \cdots, u(t - \tau_k), \dot{u}(t - \tau_k)) \tag{14}$$

where t and τ_k represent the time and time delays in the system. An approximate solution is assumed of the form:

$$u(t, \theta)_N = \begin{cases} \chi(t) + \xi(t)N(t, \theta) & \text{if } t \geq t_0 \\ \chi(t) & \text{if } t < t_0 \end{cases} \tag{15}$$

Given the inputs $X = \{x_1, x_2, \cdots, x_k\}$, weights $W = \{w_1, w_2, \cdots, w_q\}$ and the biases $B = \{b_1, b_2, \cdots, b_k\}$ such that $W, B \in \theta$ where θ is the parameter of the neural network $N(t, \theta)$. The first and second-order derivatives of the approximate solution in (2) are given by:

$$\dot{u}(t, \theta_N) = \begin{cases} \dot{\chi}(t) + \xi(t)\dot{N}(t, \theta) + \dot{\xi}(t)N(t, \theta) & \text{if } t \geq t_0 \\ \dot{\chi}(t) & \text{if } t < t_0 \end{cases} \tag{16}$$

$$\ddot{u}(t, \theta_N) = \begin{cases} \ddot{\chi}(t) + \xi(t)\ddot{N}(t, \theta) + 2\dot{\xi}(t)\dot{N}(t, \theta) + \ddot{\xi}(t)N(t, \theta) & \text{if } t \geq t_0 \\ \ddot{\chi}(t) & \text{if } t < t_0 \end{cases} \tag{17}$$

The assumed function $\xi(t)$ is chosen in such a manner that (15) approximates the solution to (14). Given the use of three hidden layers, each containing ten nodes, $N(t, \theta)$ is developed as follows:

$$h_i = x_i w + b_i \tag{18}$$

$$h_{activated} = \sigma(x_i w + b_i) \tag{19}$$

where v is either a sigmoid function:

$$sig(t) = \frac{1}{1 + e^{-t}} \tag{20}$$

or a ReLU function:

$$ReLU(t) = \max(0, t) = \begin{cases} 0 & \text{if } t \leq 0 \\ t & \text{if } t > 0 \end{cases} \tag{21}$$

Extending to all layers, let the weights be given as I, layer 1 output is given by:

$$N^{(1)}(t, \theta) = \sigma(W_1 X + b_1) \tag{22}$$

Layer 2 output is

$$N^{(2)}(t, \theta) = \sigma(W_2 N^{(1)}(t, \theta) + b_2) \tag{23}$$

Layer 3 output becomes

$$N^{(3)}(t, \theta) = \sigma(W_3 N^{(2)}(t, \theta) + b_3) \tag{24}$$

The final network output is expressed as:

$$N(t, \theta) = w_4 N^{(3)}(t, \theta) \tag{25}$$

The first architecture, denoted as $N(t, \theta)_1$, incorporates three hidden layers, each utilizing the ReLU (Rectified Linear Unit) activation function. The schematic diagram for this architecture is given in Fig. 2.

Likewise, the second architecture represented as $N(t, \theta)_2$, is also structured with three hidden layers, but employs a series of sigmoid activation functions. This is given in Fig. 3.

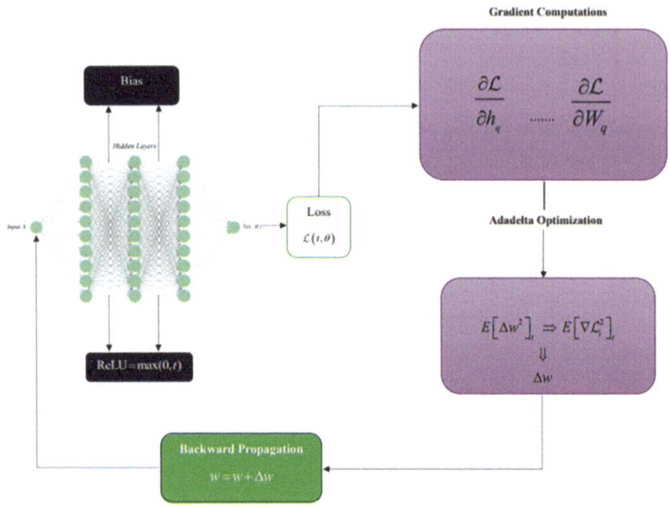

Fig. 2 Schematic diagram for architecture 1 ($SODDE_1$)

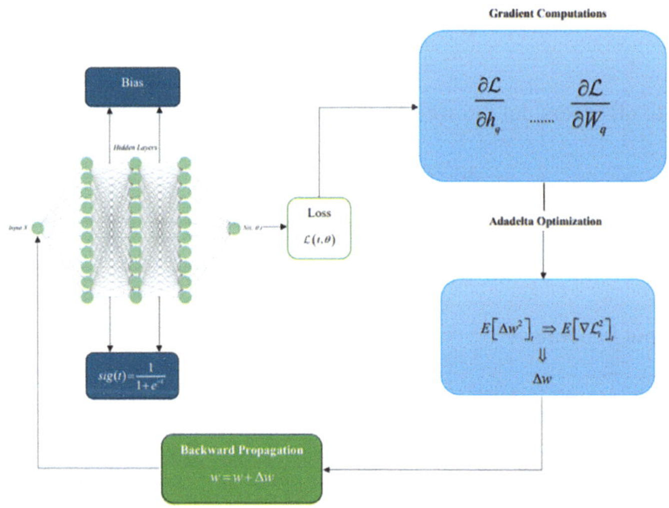

Fig. 3 Schematic diagram for architecture 2 ($SODDE_2$)

3.3 Optimization Process

Subsequently, the parameters within the architecture are optimized using the Adadelta stochastic gradient descent (SGD) method. The optimization process is formulated as the minimization of the Mean Squared Error (MSE), given by:

$$MSE = \frac{1}{2}\left[\hat{u}j(t, C) - uj\right]^2 \tag{26}$$

Expanding (30), the loss function to be optimized is given as:

$$L(t, C) = \frac{1}{2}\left[\left\{\ddot{X}(t) + 2\xi(t)\ddot{N}(t, \theta) + 2\dot{\xi}(t)\dot{N}(t, \theta) + \ddot{\xi}(t)N(t, \theta)\right\} - \ddot{u}\right]^2 \tag{27}$$

The optimization process employs the Adadelta algorithm, selected for its ability to adaptively adjust learning rates based on the running averages of squared gradients. The update rule for any parameter θ is expressed as:

$$\Delta\theta_t = -\frac{\sqrt{E[\Delta\theta^2]_{t-1} + \hat{O}}}{\sqrt{E[g(\theta)^2]t + \hat{O}}} g(\theta)_t \tag{28}$$

In this case, $g(\theta)_t$ represents the gradient of θ at time t. To facilitate the optimization process, the running averages of the squared gradients and parameter updates are computed as follows:

$$E[g(\theta)^2]_t = \rho E[g(\theta)^2]_{t-1} + (1-\rho)g(\theta)_t^2 \tag{29}$$

$$E[\Delta\theta^2]t = \rho E[\Delta\theta^2]_{t-1} + (1-\rho)(\Delta\theta_t)^2 \tag{30}$$

These calculations help ensure that the learning process remains stable and converges effectively. Using the decay constant ρ and the stability constant $\hat{O}$, the gradients are computed through backpropagation. This approach enables the model to learn from errors in previous iterations, progressively refining its predictions.

In this context, for any layer q, the preliminary gradients of the weights W is computed as:

$$h_q = t_{q-1}W_q + b_q \tag{31}$$

$$z_q = \sigma(h_q) \tag{32}$$

where:
h_q is the linear combination before the activation,
t_{q-1} is the output from the previous layer (or input for the first layer),
W_q are the weights of layer,
b_q is the bias of layer (omitted in the last layer).

The backward propagation using the Adadelta algorithm begins with the computation of the gradient of the loss L with respect to the pre-activation h_q at the output layer:

$$\frac{\partial L}{\partial h_q} = \frac{\partial L}{\partial \ddot{u}} \odot \sigma'(h_q), \tag{33}$$

where $\ddot{u} = z_j$ and $\odot$ denotes element-wise multiplication. For the weight matrix W_q at the output layer:

$$\frac{\partial L}{\partial W_q} = z_{q-1}^T \cdot \frac{\partial L}{\partial h_q} \tag{34}$$

The gradient of the loss L with respect to the pre-activation h_q at a hidden layer is:

$$\frac{\partial L}{\partial h_q} = \left[\frac{\partial L}{\partial h_{q+1}} W_{q+1}^T\right] \odot \sigma'(h_q). \tag{35}$$

For the weight matrix W_q at hidden layer q:

$$\frac{\partial L}{\partial W_q} = z_{q-1}^T \cdot \frac{\partial L}{\partial h_q} \tag{36}$$

Let $w \in W$ be a weight in the set of weights W. The accumulation of squared gradients and the accumulation of squared updates is computed as:

$$E[\Delta w^2]_t = \rho E[\Delta w^2]_{t-1} + (1-\rho)\Delta w^2 \tag{37}$$

$$E[\nabla L_i^2]_t = \rho E[\nabla L_i^2]_{t-1} + (1-\rho)\left(\frac{\partial L}{\partial w}\right)^2 \tag{38}$$

where $\rho \in (0, 1)$ is the decay rate. The adaptive learning rate is given by:

$$\Delta w = \frac{\sqrt{E[\Delta w^2]_{t-1} + \hat{O}}}{\sqrt{E[\nabla L_i^2]_t + \hat{O}}} \cdot \frac{\partial L}{\partial w} \tag{39}$$

where $\epsilon > 0$ is a small constant, typically 10^{-6}. Hence, for each layer q, the accumulation of squared gradients and the accumulation of squared updates is computed as described above.

$$E[\Delta w^2]_t = pE\left[\left(-\frac{\sqrt{E[\Delta w^2]_{t-1} + \hat{O}}}{\sqrt{E[\nabla L_i^2] + \hat{O}}} \cdot \frac{\partial L}{\partial w}\right)^2\right]_{t-1} + (1-p)\left[\left(-\frac{\sqrt{E[\Delta w^2]_{t-1} + \hat{O}}}{\sqrt{E[\nabla L_i^2]_t + \hat{O}}} \cdot \frac{\partial L}{\partial w}\right)^2\right]$$

$$E[\nabla L_i^2]_t = pE[\nabla L_i^2]_{t-1} + (1-p)\left(z_{q-1}^T \cdot \frac{\partial L_i}{\partial h_q}\right)^2 \tag{40}$$

Fig. 4 Optimization process for the proposed technique

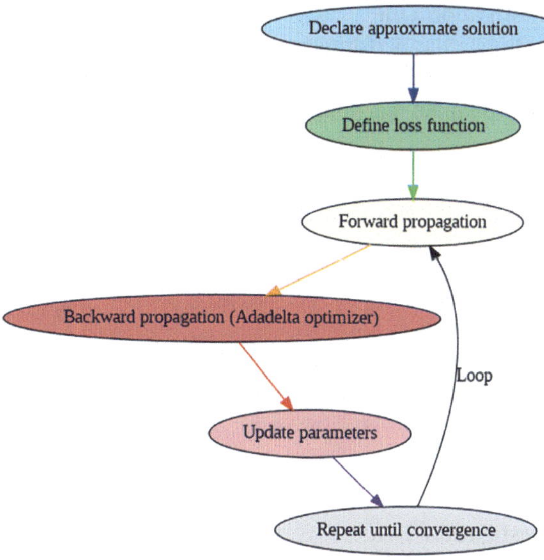

Finally, the update rule for all parameters is given as:

$$\theta = \theta + \Delta\theta \tag{41}$$

Equation 41 is used in the training process to update parameters until the error is minimized. The complete training process is illustrated in Fig. 4. The optimization process of the proposed technique was carried out using PyTorch on a Google Compute Engine instance with 12.7 GB of RAM, a 2.20 GHz processor, and a Python 3 environment. This configuration provided the advantage of a portable system for training the model without requiring complex hardware or software. The neural network parameters were updated using the Adadelta optimization algorithm, which adaptively adjusts the learning rate based on the norm of the gradient.

3.4 Application of the Proposed Technique

These investigations are conducted to demonstrate the efficiency and accuracy of the proposed methods for solving second-order Delay Differential Equations (DDEs). The models are trained over time values prior to t_0, enabling them to effectively capture the system's behavior. Subsequent sections of this paper present an analysis of these results in comparison with the Runge–Kutta and Euler methods.

Investigation One

Investigate the performance of the proposed architectures in solving the following Second-Order Delay Differential Equation (SODDE), with a focus on their ability to approximate the solution accurately and efficiently in the presence of delays.

$$\begin{aligned} \ddot{y}(t) &= y(t-\pi) & 0 \le t \le \pi \\ y(t) &= sin(t) & -\pi \le t \le 0 \end{aligned}$$

subject to:

$$y(0) = 0 \qquad \dot{y}(0) = 1$$

exact

$$y(t) = sin(t)$$

Investigation Two

Given the following SODDE, the proposed architectures are applied to find the solution of the given SODDE.

$$\begin{aligned} \ddot{y}(t) = -\frac{1}{2}y(t) + \frac{1}{2}y(t-\pi), & \qquad t \in [0, \pi] \\ y(t) = 1 - sin(t) & \qquad -\pi \le t \le 0 \end{aligned}$$

subject to:

$$y(0) = 1 \qquad \dot{y}(0) = -1$$

exact

$$y(t) = 1 - sin(t)$$

Investigation Three

Solve the following SODDE and compare the obtained solution with the exact values, as well as with the results from the fourth-order Runge–Kutta (RK4) and Euler methods. Analyze the differences in accuracy and computational efficiency among the methods to evaluate their performance.

$$\begin{aligned} \ddot{y}(t) = -\frac{sin(t)}{2 - sin(t)}y(t-\pi), & \qquad 0 \le t \le \pi \\ y(t) = 2 + sin(t) & \qquad -\pi \le t \le 0 \end{aligned}$$

subject to:

$$y(0) = 2 \qquad \dot{y}(0) = -1$$

Investigation Four
Provide the solution for the following second-order Delay Differential Equation

$$\ddot{y}(t) = -\frac{1}{2}y(t) + \frac{1}{2}y(t-\pi), \qquad t \in [0, -8\pi]$$
$$y(t) = sin(t) \qquad -\pi \leq t \leq 0$$

subject to:

$$y(0) = 0 \qquad \dot{y}(0) = 1$$

4 Findings

The Findings section presents the results of four investigations, showing the absolute errors for the proposed ANN method, the fourth-order Runge–Kutta (RK4) method, and the Euler method in solving second-order delay differential equations. The following tables provide a detailed comparison between the exact and approximate results.

4.1 Training Loss

Generally, the training loss of the models follows a similar pattern, exhibiting an oscillatory reduction of error that approaches zero as the number of training epochs increases. The choice of training data depends on the problem; however, the general practice is to train the model on historical values, i.e., at $t < 0$. The two architectures perform differently for each problem, and their optimization errors are presented in the Figs. 5, 6, 7 and 8.

4.2 Accuracy Comparison

Using proposed architectures, the trained ANN model was tested on investigation one at a time interval spanning from $t = 0.0$ to $t = 10.0$. Using the proposed architectures, the accuracy of the model is experimented with using investigation two spanning from $t = 0.0$ to $t = 10.0$. For investigation 3, the proposed architectures were utilized. Investigation Three spans from $t = 0.0$ to $t = 8.0$. Lastly, Investigation Four spans from $t = 10.0$ to $t = 20.0$ using the proposed architectures (Tables 1, 2 and 3).

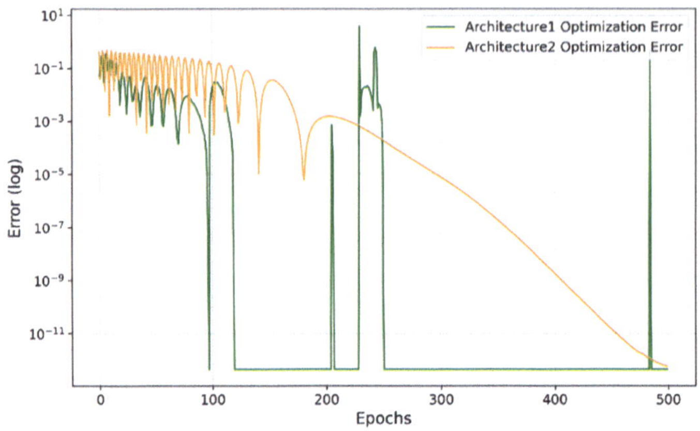

Fig. 5 Optimization error for solving investigation one using the proposed architectures

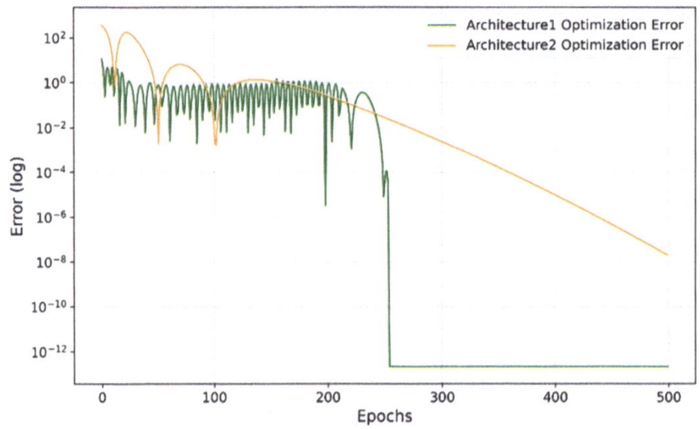

Fig. 6 Optimization error for solving investigation two using the proposed architectures

4.3 *Error Analysis*

This section investigates the efficiency of the proposed ANN-based solution in comparison with the numerical methods Runge–Kutta and Euler. The degree of accu-racy of these methods for approximating solutions of second-order delay differential equations (SODDEs) is analyzed by comparing the absolute errors at different time steps. The analysis provides insight into error behavior over time and demonstrates how the ANN approach effectively reduces such errors, particularly when handling complex delayed dynamics that are difficult to manage using traditional numerical techniques (Tables 4 and 5).

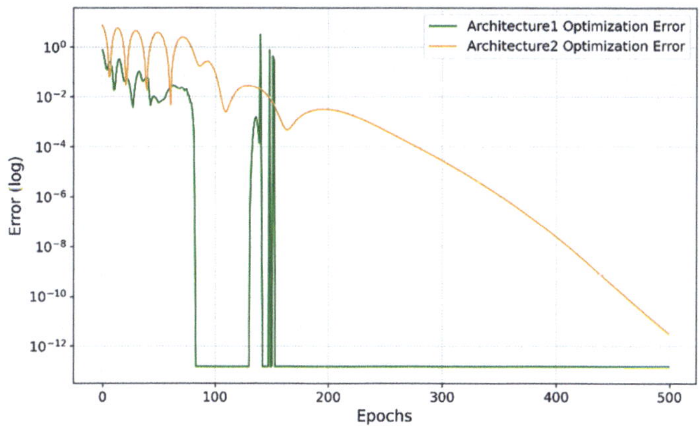

Fig. 7 Optimization error for solving investigation three using the proposed architectures

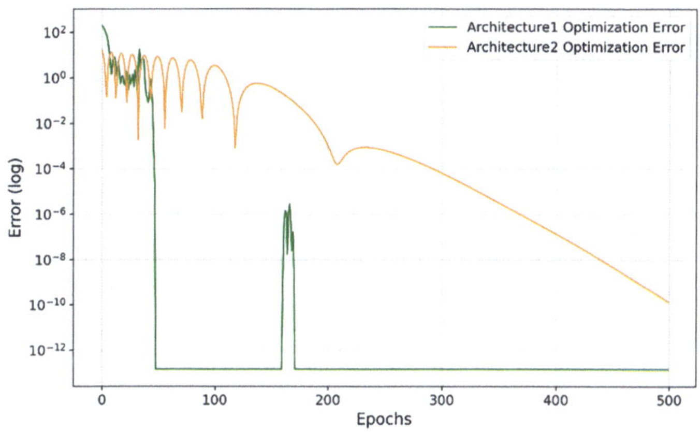

Fig. 8 Optimization error for solving investigation four using the proposed architectures

5 Findings Discussion

The research results demonstrate that the proposed ANN_{SODDE} method achieves outstanding performance through its ReLU (Architecture 1) and Sigmoid (Architecture 2) variants when solving Second-Order Delay Differential Equations (SODDEs). Benchmark problems indicate that both architectures outperform the fourth-order Runge–Kutta (RK4) and Euler methods. Notably, the absolute error for the ReLU-based architecture reaches zero in the investigations. The ANN approach exhibits inherent accuracy in capturing complex functional relationships in SODDEs, regardless of the activation function used in these architectures, making it a reliable alternative to traditional numerical methods. Direct performance comparisons reveal nearly identical numerical outcomes between the two architectures, as shown in the absolute

Table 1 Exact and approximate results for investigation two

Time (t)	Exact	$Proposed ANN_{SODDE1}$	$Proposed ANN_{SODDE2}$	Runge–Kutta	Euler
0.0	0.000000000	0.000000000	0.000000000	0.000000000	0.000000000
1.0	0.84147100	0.84147100	0.84147100	0.84147100	0.84152100
2.0	0.90929700	0.90929700	0.90929700	0.90929700	0.90945200
3.0	0.14112000	0.14112000	0.14112000	0.14112000	0.14133800
3.0	0.14112000	0.14112000	0.14112000	0.14112000	0.14133800
4.0	−0.75680200	−0.75680200	−0.75680200	−0.75680200	−0.75661900
5.0	−0.95892400	−0.95892400	−0.95892400	−0.95892400	−0.95879700
6.0	−0.27941500	−0.27941500	−0.27941500	−0.27941500	−0.27917900
7.0	0.65698700	0.65698700	0.65698700	0.65698700	0.65763600
8.0	0.98935800	0.98935800	0.98935800	0.98935800	0.99068400
9.0	0.41211800	0.41211800	0.41211800	0.41211800	0.41424800
10.0	−0.54402100	−0.54402100	−0.54402100	−0.54402100	−0.54095000

Table 2 Exact and approximate results for investigation two

Time (t)	Exact	$Proposed ANN_{SODDE1}$	$Proposed ANN_{SODDE2}$	Runge–Kutta	Euler
0.0	1.000000000	1.000000000	1.000000000	1.000000000	1.000000000
1.0	0.158529000	0.158529000	0.158533000	0.158472000	0.158472000
2.0	0.090703000	0.090703000	0.090719000	0.090554000	0.090554000
3.0	0.858880000	0.858880000	0.858917000	0.858759000	0.858759000
4.0	1.756803000	1.756803000	1.756869000	1.756880000	1.756880000
5.0	1.958924000	1.958924000	1.959027000	1.959210000	1.959210000
6.0	1.279415000	1.279415000	1.279564000	1.279680000	1.279680000
7.0	0.343013000	0.343013000	0.343216000	0.342964000	0.342964000
8.0	0.010642000	0.010642000	0.010906000	0.010235000	0.010235000
9.0	0.587882000	0.587882000	0.588216000	0.587448000	0.587448000
10.0	1.544021000	1.544021000	1.544434000	1.544000000	1.544000000

Table 3 Exact and approximate results for investigation three

Time (t)	Exact	$Proposed ANN_{SODDE1}$	$Proposed ANN_{SODDE2}$	Runge–Kutta	Euler
0.0	2.000000000	2.000000000	2.000000000	2.000000000	2.000000000
1.0	2.841471000	2.841471000	2.841471000	2.841360000	2.841410000
2.0	2.909297000	2.909297000	2.909297000	2.908750000	2.908920000
3.0	2.141120000	2.141120000	2.141120000	2.140410000	2.140650000
4.0	1.243197000	1.243197000	1.243197000	1.242430000	1.242630000
5.0	1.041076000	1.041076000	1.041076000	1.040060000	1.040160000
6.0	1.720585000	1.720585000	1.720584000	1.719270000	1.719350000
7.0	2.656987000	2.656987000	2.656986000	2.655400000	2.655580000
8.0	2.989358000	2.989358000	2.989357000	2.987540000	2.987820000

Table 4 Exact and approximate results for investigation four

Time (t)	$Exact_{SODDE}$	$ProposedANN_{SODDE1}$	$ProposedANN_{SODDE2}$	$RK4_{SODDE1}$	$Euler_{SODDE1}$
10.0	−0.544021000	−0.544021000	−0.544001000	−0.544021000	10.000000000
11.0	−0.999990000	−0.999990000	−0.999965000	−0.999990000	−1.000420000
12.0	−0.536573000	−0.536573000	−0.536543000	−0.536573000	−0.537110000
13.0	0.420167000	0.420167000	0.420202000	0.420167000	0.420055000
14.0	0.990607000	0.990607000	0.990648000	0.990607000	0.991118000
15.0	0.650288000	0.650288000	0.650334000	0.650288000	0.651022000
16.0	−0.287903000	−0.287903000	−0.287851000	−0.287903000	−0.287633000
17.0	−0.961397000	−0.961397000	−0.961338000	−0.961397000	−0.961921000
18.0	−0.750987000	−0.750987000	−0.750920000	−0.750987000	−0.751893000
19.0	0.149877000	0.149877000	0.149952000	0.149877000	0.149432000
20.0	0.912945000	0.912945000	0.913028000	0.912945000	0.913454000

Table 5 Absolute errors for investigation one

Time (t)	$ProposedANN_{SODDE1}$	$ProposedANN_{SODDE2}$	Runge–Kutta	Euler
0.0	0.000000000E+00	0.000000000E+00	0.000000000E+00	0.000000000E+00
1.0	0.000000000E+00	0.000000000E+00	5.960464000E-08	5.000000000E-05
2.0	0.000000000E+00	0.000000000E+00	4.172325000E-07	1.550000000E-04
3.0	0.000000000E+00	2.980232000E-08	0.000000000E+00	2.180000000E-04
4.0	0.000000000E+00	5.960464000E-08	4.768372000E-07	1.840000000E-04
5.0	0.000000000E+00	5.960464000E-08	2.980232000E-07	1.270000000E-04
6.0	0.000000000E+00	5.960464000E-08	4.768372000E-07	2.360000000E-04
7.0	0.000000000E+00	5.960464000E-08	4.172325000E-07	6.490000000E-04
8.0	0.000000000E+00	5.960464000E-08	2.384186000E-07	1.326000000E-03
9.0	0.000000000E+00	8.940697000E-08	5.066395000E-07	2.129000000E-03
10.0	0.000000000E+00	1.192093000E-07	1.192093000E-07	3.071000000E-03

error tables. However, the optimization error data in Figs. 5, 6, 7 and 8 indicate minor variations between the ReLU-based (Architecture 1) and Sigmoid-based (Architecture 2) ANN_{SODDE} models, and their final solution precision is not equivalent (Figs. 9, 10, 11 and 12).

These results suggest that the choice of activation function affects the level of precision achieved in solving SODDE problems. Simulation times further highlight error accumulation in both the Runge–Kutta 4th-order and Euler methods, whereas the ANN-based methods show no signs of such accumulation. This underscores a limitation of traditional numerical techniques for SODDEs, where error propagation can significantly reduce solution accuracy over longer time horizons. The ANN_{SODDE} architectures, by contrast, maintain accuracy consistently, demonstrating their robustness. Overall, the findings establish the ANN_{SODDE} method with ReLU as a highly promising and robust approach for solving SODDEs, surpassing traditional numerical techniques and offering a new paradigm for tackling challenging SODDEs and delay differential equations more broadly (Tables 6, 7 and 8).

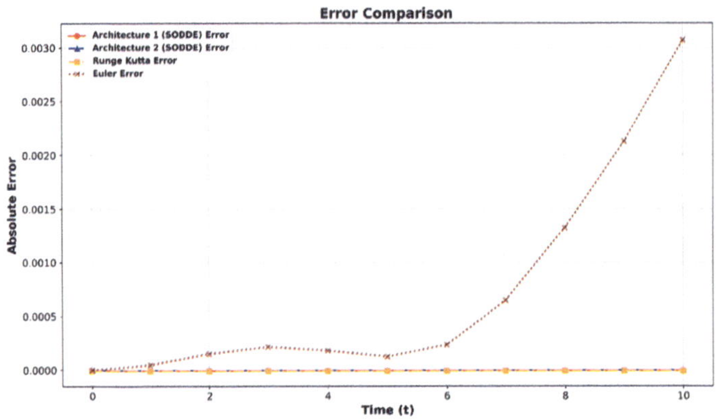

Fig. 9 Error comparison of the proposed SODDE methods, analytical solution, and numerical methods (RK4 and Euler) for investigation one

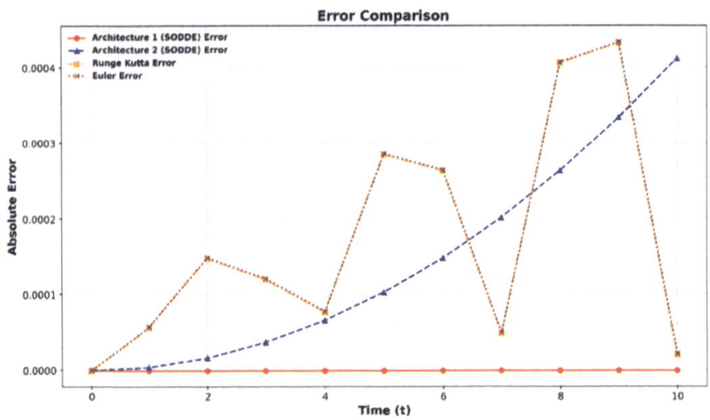

Fig. 10 Error comparison of the proposed SODDE methods, analytical solution, and numerical methods (RK4 and Euler) for investigation two

6 Conclusion

This research makes significant contributions to computational mathematics through the development of ANN_{SODDE}, which demonstrates superior performance as an Artificial Neural Network (ANN)-based method for solving Second-Order Delay Differential Equations (SODDEs) using ReLU and Sigmoid activation functions. The pure MLP-ANN framework incorporates the history function as an inherent element, making both architectures a superior alternative to traditional numerical approaches. The investigations conclusively show that ANN_{SODDE} architectures with ReLU-based (Architecture 1) or Sigmoid-based (Architecture 2) activation functions outperform the fourth-order Runge–Kutta (RK4) and Euler methods in terms of accuracy for the tested SODDE problems. Both ANN architectures exhibit

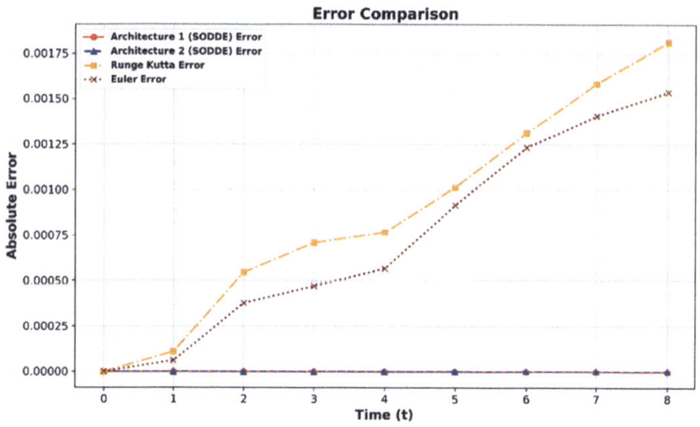

Fig. 11 Error comparison of the proposed SODDE methods, analytical solution, and numerical methods (RK4 and Euler) for investigation three

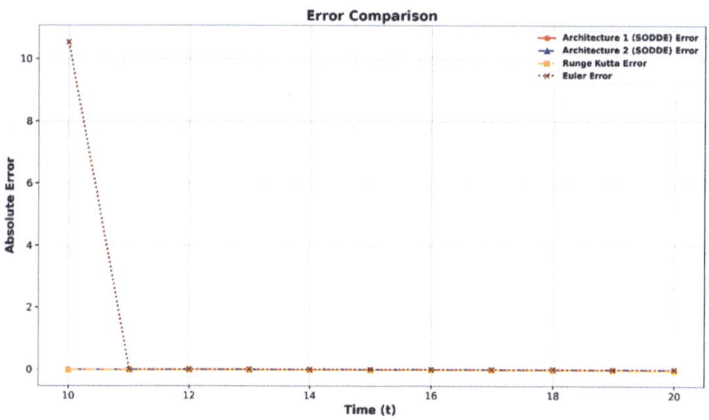

Fig. 12 Error comparison of the proposed ANN_{SODDE1} method, analytical solution, and numerical methods (RK4 and Euler) for investigation four

clear superiority over traditional numerical methods, delivering near-zero absolute errors throughout their computations, thereby demonstrating remarkable precision and reliability. The results indicate that both ReLU and Sigmoid activation functions achieve similarly high levels of accuracy for the studied SODDEs with the given network configurations, suggesting that accuracy is largely independent of the choice between ReLU and Sigmoid in the hidden layers. Both ANN_{SODDE} architectures combine exceptional accuracy with robust computational efficiency, making ANN-based solutions an attractive alternative to conventional numerical methods. This approach provides researchers and practitioners with a valuable tool for modeling complex delayed dynamic systems, offering reduced error propagation and potentially lower computational burden.

Table 6 Absolute errors for investigation two

Time (t)	$ProposedANN_{SODDE1}$	$ProposedANN_{SODDE2}$	Runge–Kutta	Euler
0.0	0.000000000E+00	0.000000000E+00	0.000000000E+00	0.000000000E+00
1.0	0.000000000E+00	4.000000000E-06	5.700000000E-05	5.700000000E-05
2.0	0.000000000E+00	1.700000000E-05	1.490000000E-04	1.490000000E-04
3.0	0.000000000E+00	3.700000000E-05	1.210000000E-04	1.210000000E-04
4.0	0.000000000E+00	6.600000000E-05	7.700000000E-05	7.700000000E-05
5.0	0.000000000E+00	1.030000000E-04	2.860000000E-04	2.860000000E-04
6.0	0.000000000E+00	1.490000000E-04	2.650000000E-04	2.650000000E-04
7.0	0.000000000E+00	2.020000000E-04	4.900000000E-05	4.900000000E-05
8.0	0.000000000E+00	2.640000000E-04	4.070000000E-04	4.070000000E-04
9.0	0.000000000E+00	3.340000000E-04	4.340000000E-04	4.340000000E-04
10.0	0.000000000E+00	4.130000000E-04	2.100000000E-05	2.100000000E-05

Table 7 Absolute errors for investigation three

Time (t)	$ProposedANN_{SODDE1}$	$ProposedANN_{SODDE2}$	Runge Kutta	Euler
0.0	0.000000000E+00	0.000000000E+00	0.000000000E+00	0.000000000E+00
1.0	0.000000000E+00	0.000000000E+00	1.110000000E-04	6.100000000E-05
2.0	0.000000000E+00	0.000000000E+00	5.470000000E-04	3.770000000E-04
3.0	0.000000000E+00	0.000000000E+00	7.100000000E-04	4.700000000E-04
4.0	0.000000000E+00	0.000000000E+00	7.670000000E-04	5.670000000E-04
5.0	0.000000000E+00	1.192093000E-07	1.016000000E-03	9.160000000E-04
6.0	0.000000000E+00	4.768372000E-07	1.315000000E-03	1.235000000E-03
7.0	0.000000000E+00	4.768372000E-07	1.587000000E-03	1.407000000E-03
8.0	0.000000000E+00	7.152557000E-07	1.818000000E-03	1.538000000E-03

Table 8 Absolute errors for investigation three

Time (t)	$ProposedANN_{SODDE1}$	$ProposedANN_{SODDE2}$	Runge Kutta	Euler
10.0	0.000000000E+00	2.100000000E-05	1.192093000E-07	1.054402200E+01
11.0	0.000000000E+00	2.500000000E-05	2.384186000E-07	4.300000000E-04
12.0	0.000000000E+00	3.000000000E-05	5.960464000E-08	5.370000000E-04
13.0	0.000000000E+00	3.500000000E-05	2.980232000E-08	1.120000000E-04
14.0	0.000000000E+00	4.000000000E-05	3.576279000E-07	5.110000000E-04
15.0	0.000000000E+00	4.600000000E-05	1.192093000E-07	7.340000000E-04
16.0	0.000000000E+00	5.300000000E-05	2.980232000E-07	2.700000000E-04
17.0	0.000000000E+00	6.000000000E-05	4.768372000E-07	5.240000000E-04
18.0	0.000000000E+00	6.700000000E-05	2.384186000E-07	9.060000000E-04
19.0	0.000000000E+00	7.400000000E-05	2.086163000E-07	4.450000000E-04
20.0	0.000000000E+00	8.200000000E-05	2.980232000E-07	5.090000000E-04

Competing Interest The authors have no conflicts of interest to declare that are relevant to the content of this chapter.

References

1. Ames, W.F.: Differential equations. In: Mathematics for Mechanical Engineers, pp. 5–12. CRC Press (2022)
2. Yakovlev, B.V., Nikiforova, L.V.: Mathematical modeling of a particle motion in mineral processing devices. In: 9th International Conference on Mathematical Modeling (2021)
3. Xie, W.: The fluctuation effect of international financial market based on MSV model. In: The Sixth International Conference on Information Management and Technology (2021)
4. Podlesny, S.: Mathematical modeling of typical electrical circuits, oscillators and pendulums. InterConf **26**(129), 358–373 (2022)
5. Yan, X., Bauer, R., Koch, G., Schropp, J., Perez Ruixo, J.J., Krzyzanski, W.: Delay differential equations-based models in NONMEM. J. Pharmacokinet Pharmacodyn. **48**(6), 763–802 (2021)
6. Chen-Charpentier, B.M.: On population models with delays and dependence on past values, MDPI (2024)
7. Li, W., Zhang, Z., Ge, S.S.: Stabilization control for strict-feedback nonlinear systems with time delays. IEEE Trans. Syst. Man Cybern. Syst. **52**(12), 7549–7560 (2022)
8. Hussein, A., Qasem, A.: Solving second-order ordinary differential equations by finite difference, Adomian decomposition, and Galerkin methods, vol. 7, Full Text Book of Minar Congress (2023)
9. Mutuk, H.: Solutions of second-order linear differential equations via Runge-Kutta, power series, and Laplace transformation in Mathematica. Iqdır Universitesi Fen Bilimleri Enstitusu Dergisi **9**(4), 2048–2061 (2019)
10. Sabir, Z., Wahab, H.A., Umar, M., Erdoğan, F.: Stochastic numerical approach for solving second-order nonlinear singular functional differential equation. Appl. Math. Comput. **363**, 124–605 (2019)
11. Khan, I., Raja, M.A.Z., Shoaib, M., Kumam, P., Alrabaiah, H., Shah, Z., Islam, S.: Design of neural network with Levenberg–Marquardt and Bayesian regularization backpropagation for solving pantograph delay differential equations. IEEE Access **8**, 137918–137933 (2020)
12. Regazzoni, F., Dedè, L., Quarteroni, A.: Machine learning for fast and reliable solution of time-dependent differential equations. J. Comput. Phys. **397**, 108–852 (2019)
13. Audu, K.J., Benjamin, M., Mohammed, U., Yahaya, Y.A.: Solving ordinary differential equations of first order using artificial neural network approach, Unpublished (2024)
14. Audu, K.J., Benjamin, M., Mohammed, U., Yahaya, Yahaya, Y.A.: Utilizing the artificial neural network approach for the resolution of first-order ordinary differential equations. Malays. J. Sci. Adv. Technol. **4**(3), 210–216 (2024)
15. Sivalingam, S.M., Pushpendra, K., Venkatesan, G.: A neural networks-based numerical method for the generalized Caputo-type fractional differential equations. Math. Comput. Simul. **213**, 302–323 (2023)
16. Sivalingam, S.M., Pushpendra, K., Govindaraj, V.: A novel optimization-based physics-informed neural network scheme for solving fractional differential equations. Eng. Comput. **40**(2), 1–11 (2023)
17. Sivalingam, S.M, Pushpendra, K., Govindaraj V.: Chebyshev neural network-based numerical scheme to solve distributed-order fractional differential equations. Comput. Math. Appl. **164**, 150–165 (2023)
18. Liu, Z., Yang, Y., Cai, Q.: Neural network as a function approximator and its application in solving differential equations. Appl. Math. Mech. **40**(2), 237–248 (2019)
19. Weerakody, P.B., Wong, K.W., Wang, G., Ela, W.: A review of irregular time series data handling with gated recurrent neural networks. Neurocomputing **441**, 161–178 (2021)

20. Monsel, T., Semeraro, O., Mathelin, L., Charpiat, G.: Neural State-Dependent Delay Differential Equations. Elsevier BV (2023)
21. Smagulova, K., James, A.P.: A survey on LSTM memristive neural network architectures and applications. Eur. Phys. J. Spec. Top. **228**(10), 2313–2324 (2018)
22. Schmidgall, S., Achterberg, J., Miconi, T., Kirsch, L., Ziaei, R., Hajiseyedrazi, S.P., Eshraghian, J.: Brain-inspired learning in artificial neural networks: a review (2023)
23. Ji, X., Orosz, G.: Learning time delay systems with neural ordinary differential equations. IFAC-PapersOnLine **55**(36), 79–84 (2022)
24. Salah, J.: Some propositions on mathematical models of population growth. Center for Open Science (2023)
25. Vinodbhai, C.D., Dubey, S.: Numerical solution of neutral delay differential equations using orthogonal neural network. Sci. Rep. **13**(1), 1–24 (2023)
26. Pantidis, P., Mobasher, M.E.: Integrated finite element neural network (I-FENN) for non-local continuum damage mechanics. Comput. Methods Appl. Mech. Eng. **404**, 115–766 (2023)
27. Dorodnitsyn, V.A., Kozlov, R., Meleshko, S.V., Winternitz, P.: Second-order delay ordinary differential equations, their symmetries and application to a traffic investigation. J. Phys. A: Math. Theor. **54**(10), 105–204 (2021)
28. Schmidgall, S., Achterberg, J., Miconi, T., Kirsch, L., Ziaei, R., Hajiseyedrazi, S.P., Eshraghian, J.: Brain-inspired learning in artificial neural networks: a review (2023)
29. Khanbehbin, T., Effati, S., Gachpazan, M., Miri, S.: A new approach for solving delay differential equations with time-varying delay. Int. J. Nonlinear Anal. Appl. **15**(10), 235–241 (2023)
30. Salah, J.: Some propositions on mathematical models of population growth. Center for Open Science (2023)
31. Ji, Z.: Approximation of continuous functions by artificial neural networks. Honors Thesis, Union College (2018)
32. Rasdi, N., Majid, Z.A., Ismail, F., Senu, N., See, P.P., Radzi, H.M.: Solving second order delay differential equations by direct two- and three-point one-step block method. Appl. Math. Sci. **7**(54), 2647–2660 (2013)
33. Xuan, Y., Amin, R., Zaman, F., Khan, Z., Ullah, I., Nazir, S.: Second-order delay differential equations to deal the experimentation of internet of industrial things via Haar wavelet approach. Discrete Dyn. Nat. Soc. (2021)
34. Aljawia, S., Khan, K., Hussain, S., Shah, K., Abdeljawad, T.: On the numerical solution of second order delay differential equations via a novel approach. J. Math. Comput. Sci. **36**, 1–16 (2025)
35. He, J., Tiamiyu, A.T.: Physics-informed neural networks in iterative form of nonlinear equations for numerical algorithms and simulations of delay differential equations. XXPhys. A **660**(2025), 130–368 (2025)

Employing Shifted Chebyshev Polynomial Techniques within Neural Network Architectures for Solving Damped Harmonic Oscillator Equations

Sabastine Emmanuel, Saratha Sathasivam, Majid Khan Majahar Ali, and Abdulwaheed Salaudeen Adebayo

Abstract The precise resolution of nonlinear differential equations is essential for representing intricate occurrences in science and engineering. Conventional numerical methods frequently exhibit computational inefficiency and diminished accuracy, especially when utilized for high-dimensional systems with complex boundary conditions. This chapter introduces the Shifted Chebyshev Neural Network (SCNN) as a resilient option for enhancing both precision and efficiency in addressing these equations. By incorporating Chebyshev polynomials into a neural network framework, the SCNN adeptly captures nonlinear system dynamics and improves convergence characteristics. Comparative assessments with conventional numerical methods reveal that SCNNs provide more accurate approximations of the precise solution while decreasing computational expenses. The results indicate that SCNNs present a reliable and scalable framework for addressing differential equations, hence creating new possibilities for sophisticated modeling in fields such as engineering, physics, and applied mathematics. This contribution establishes a foundation for enhanced analytical accuracy and expanded scientific innovation.

S. Emmanuel
Department of Mathematics, Federal University Lokoja, Lokoja, Nigeria
e-mail: sabastine.emmanuel@fulokoja.edu.ng

S. Sathasivam (✉) · M. K. Majahar Ali
School of Mathematical Sciences, Universiti Sains Malaysia, George Town, Penang, Malaysia
e-mail: saratha@usm.my

M. K. Majahar Ali
e-mail: majidkhanmajaharali@usm.my

A. S. Adebayo
Saadu Zungur University, Bauchi, Nigeria
e-mail: salaudeenadebayo@sazu.edu.ng

L. Kaur and P. Kumar (eds.), *Neural Networks: Theory, Algorithms, Simulation, and Applications*, Studies in Computational Intelligence 1273,
https://doi.org/10.1007/978-3-032-18750-5_7

1 Introduction

Differential equations form the backbone of scientific models across numerous fields, describing the dynamic behaviour of various systems in physics, engineering, biology, and beyond. Solving these equations is pivotal in understanding and predicting the evolution of physical phenomena. Computational modelling of differential equations governed by dynamical systems is a challenging problem in many scientific fields, such as computational biology, engineering, and physics [1]. Traditional numerical techniques often face limitations in accurately capturing the behaviour of complex systems over extended time frames [2]. Researchers have explored innovative approaches to address these challenges, notably employing artificial neural networks (ANNs) as powerful tools for solving differential equations and modelling dynamic systems due to their capacity for function approximation and non-linear mapping [3, 4]. In order to accurately approximate and predict dynamic systems, this study will investigate the use of Shifted Chebyshev Neural Networks (SCNN), with a particular focus on the Damped Harmonic Oscillator (DHO) equation. The DHO equation represents a fundamental second-order differential equation describing systems subjected to damping forces and restoring forces.

$$\frac{d^2x}{dt^2} = f\left(t, x, \frac{dx}{dt}\right), \quad t \in [a, b] \tag{1}$$

with initial conditions: x(0) = A, x'(0) = A'.

The literature presents a rich landscape of research encompassing various aspects relevant to this study. Xu et al. [5], Raissi et al. [6] explore spectral methods and deep learning techniques for solving differential equations, providing a foundation for numerical methods and neural network applications in scientific computing. Additionally, the work by [7, 8] showcases the efficacy of neural networks in approximating solutions to ordinary and partial differential equations, demonstrating their potential in physics-informed modelling and inverse problems. The efficacy of support vector neural networks (SCNN) in solving partial differential equations and approximating complicated functions has been brought to light by recent improvements [8]. While [9–11] introduces a new method that uses neural networks to solve differential equations, [12] give a thorough analysis of Chebyshev polynomials, which are essential for neural network input space transformations [12–14]. Comprehensive research have not yet examined the generalizability of SCNNs applied to differential equations with boundary and initial condition variables. There has to be further research into the possibility of SCNNs adapting to different physical systems when the exact solution isn't known or easily available. Traditional approaches to solving complicated differential equations can suffer from slow convergence, computational complexity, and excessive memory utilization. The Shifted Chebyshev Neural Network (SCNN), on the other hand, takes advantage of ANN architecture to offer a more efficient option. Shorter convergence times for difficult problems are achieved using SCNNs, which are composed of interconnected processing units (neurones)

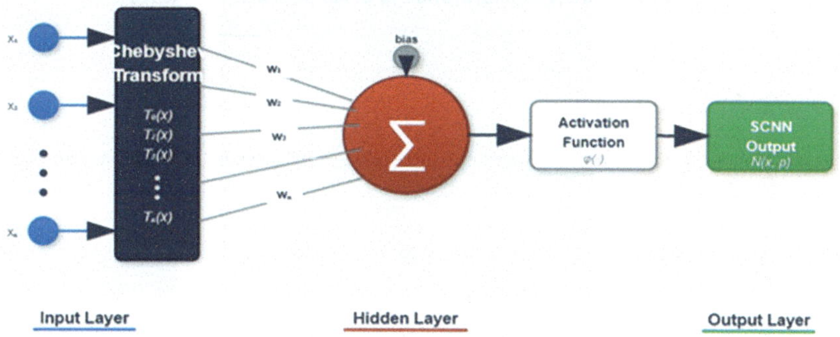

Fig. 1 Structure of the Shifted Chebyshev Neural Network (SCNN)

that do fundamental computations at each neurone and require minimum memory. Because there are so many connections between neurons, the network can store and process a lot of data, especially data related to things that people have learned or experienced. Synaptic weights, which are the strengths of connections between neurons, are a powerful tool for modeling and solving complicated differential equations, and SCNNs are no exception. This research intends to add to the current literature by investigating the feasibility of using SCNN to simulate the dynamics of the Damped Harmonic Oscillator equation. The goal is to create a powerful, efficient, and accurate prediction model for dynamic systems by using SCNN and adding shifted Chebyshev polynomials as input transformations.

1.1 Shifted Chebyshev Neural Network

Incorporating shifted Chebyshev polynomials as activation functions or input transformations is what makes the Shifted Chebyshev Neural Network (SCNN) architecture unique. Intervals are assigned to these polynomials. Boost the network's capacity to approximatively solve differential equations and complicated functions by serving as strong basis functions. A benefit of SCNN is that it uses shifted Chebyshev polynomials, which are useful for approximating functions in domains other than the typical interval of [–1, 1]. The SCNN is able to handle issues with varying ranges of values because it shifts and scales these polynomials to fit the intervals of the problem. Because of this, it can handle systems or equations that span domains. To implement SCNN, one must first provide a structured representation of the input space for neural networks by transforming the data using shifted Chebyshev polynomials. It is essential for solving differential equations, modeling dynamic systems, or accurately approximating functions that the network is able to learn and map complex relationships between inputs and outputs. This transformation makes this possible.

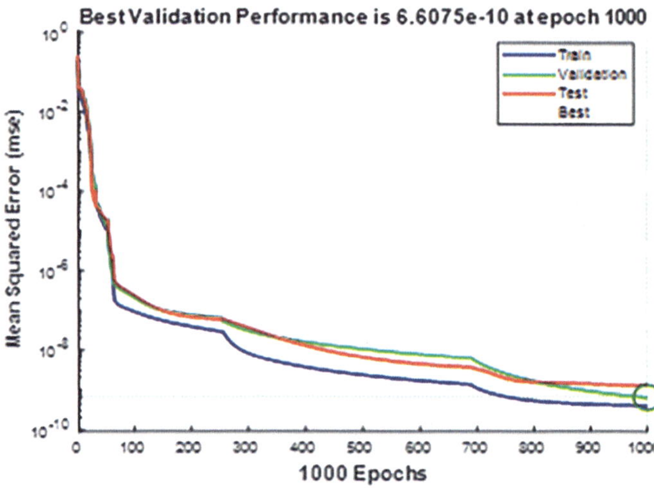

Fig. 2 Validation performance of the neural network showing convergence and prediction accuracy on the validation dataset

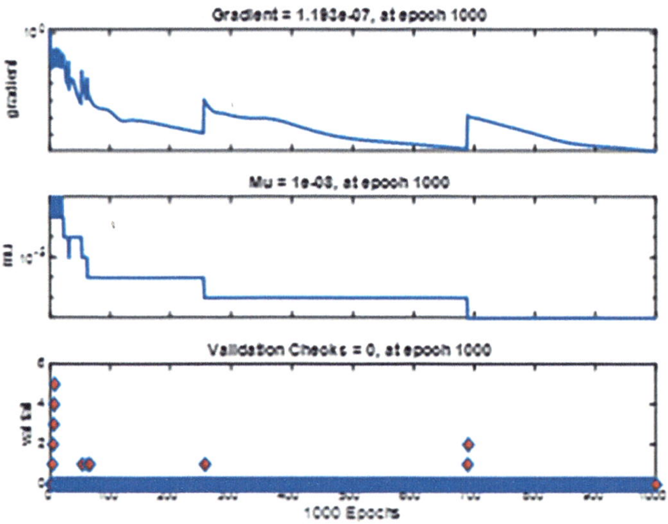

Fig. 3 Graph of gradients performance, illustrating the evolution of gradient values during training. The plot shows how the network's weights are updated over epochs, highlighting convergence behavior and the effectiveness of the gradient descent optimization in minimizing the loss function

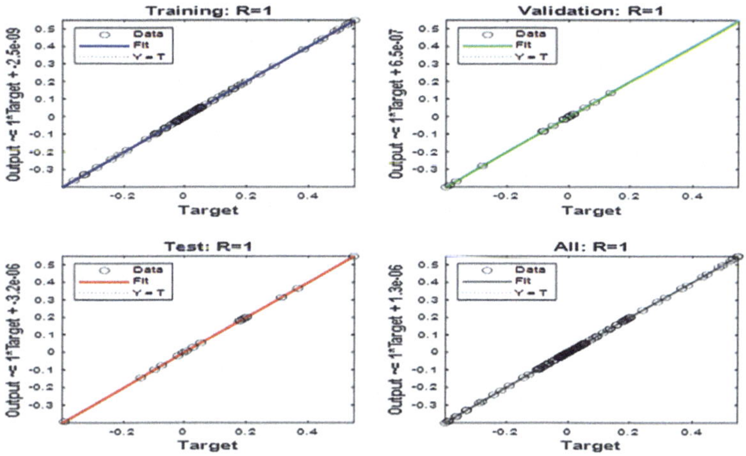

Fig. 4 Graphical representation of the regression analysis, showing the relationship between predicted and actual values. The plot demonstrates the accuracy of the model, the closeness of fit, and the effectiveness of the neural network in capturing the underlying trends in the data

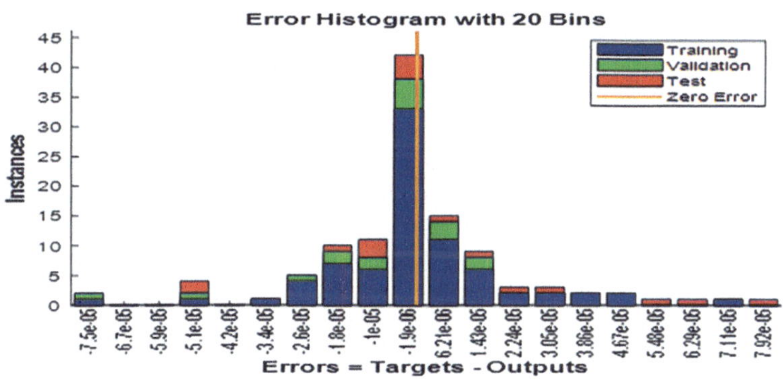

Fig. 5 Error histogram illustrating the distribution of prediction errors for the neural network

Figure 1 presents a schematic diagram of the fundamental SCNN architecture, comprising input, hidden, and output layers. Each node in these levels executes designated computations or activation functions, while the interconnections enable the transmission of information across the network. The training performance and corresponding graphical results of the neural network are presented in Figs. 2, 3, 4, and 5, offer valuable information into the performance of the model, such as: Validation Performance, it measures the model's accuracy on validation data while it is being trained. A reduction and stabilisation in error indicate a strong ability to generalise, while an increase may reflect overfitting, Regression plots evaluates the accuracy of forecasts by comparing them to the actual targets. Points that are located near the perfect correlation line, which has a slope of 1, and have a high R-value close to 1, suggest a good level of predictive accuracy and Error Histogram which displays the frequency distribution of prediction errors. A concentration close to zero signifies precise predictions, whereas outliers indicate possible problems. The methodology employs a double-layer Shifted Chebyshev Neural Network (SCNN) to tackle the present problem, aiming to model the behaviour of a Damped Harmonic Oscillator (DHO) equation. The architecture, illustrated in Fig. 1, comprises a single input node, a functional expansion block utilizing shifted Chebyshev polynomials, and a single output node. The SCNN model entails two integral components: the numerical transformation and the learning segments. Initially, within the numerical transformation phase, each input datum, denoted as $x = (x_1, x_2, x_3, \cdots, x_h)^T$, representing a single input node with h data points undergo expansion using shifted Chebyshev polynomials. These polynomials, orthogonal and derived as solutions of the shifted Chebyshev differential equations, yield a functional expansion that can be perceived as a new input vector. Specifically, the initial two Chebyshev polynomials are fundamental components in this transformed input space. The first-degree polynomial is defined as:

$$T_0(x) = 1 \tag{2}$$

$$T_1(x) = x \tag{3}$$

$$T_{k+1}(x) = 2xT_k(x) - T_{k-1}(x) \tag{4}$$

These polynomials facilitate the transformation of input data, crucial in constructing an SCNN that approximates the behaviour of the Damped Harmonic Oscillator equation. The subsequent learning segment of the SCNN involves training the network to comprehend and predict the intricate dynamics of the system based on the transformed input representations. In a neural network, specifically in the context of a Shifted Chebyshev Neural Network (SCNN) aiming to approximate the solution of a Damped Harmonic Oscillator (DHO) Equation, the weights connecting the input to the output layer are adjusted during the training process using an optimization algorithm, typically gradient descent or its variants. The adjustment of weights aims to minimize the error function, improving the network's ability to approximate the dynamics of the DHO system.

The weights $(W_(input_output))$ from the input layer to the output layer in an SCNN can be updated based on the gradient of the error function $E(\theta)$ with respect to these weights. The error function $E(\theta)$ can be formulated as the Mean Squared Residuals (MSR) or Mean Squared Error (MSE) between the left-hand side and right-hand side of the DHO Equation:

$$E(\theta) = \frac{1}{N}\sum_{i=1}^{N}\left[\frac{d^2x}{dt^2} + 2\phi\frac{dx}{dt} + \omega^2 x - f(\omega t)\right]^2 \tag{5}$$

The hyperbolic tangent function squashes the output between -1 and 1. It is commonly used in neural networks for its smoothness and suitability for modeling complex functions.

$$NN(x_i, p) = \tanh(h) \tag{6}$$

$$h = \sum_{j=1}^{N} W_j T_{j-1}(x) \tag{7}$$

Here, W_j denotes the weight vector of the Shifted Chebyshev Neural Network, and T_j represents the corresponding shifted Chebyshev polynomials.

The weight update rule using gradient descent can be expressed as:

$$W_{\text{input\_output}} = W_{\text{input\_output}} - \mu \frac{\partial E(\theta)}{\partial W_{\text{input\_output}}} \tag{8}$$

$$W_j^{(k+1)} = W_j^{(k)} - \mu \frac{\partial E^{(k)}(\theta)}{\partial W_j^{(k)}} \tag{9}$$

where:

- $W_{\text{input\_output}}$ represents the weights connecting the input layer to the output layer.
- μ denotes the learning rate, a hyperparameter determining the step size during weight updates.
- $\frac{\partial E(\theta)}{\partial W_{\text{input\_output}}}$ signifies the gradient of the error function $E(\theta)$ with respect to the weights.

The gradient $\frac{\partial E(\theta)}{\partial W_{\text{input\_output}}}$ guides the direction of weight adjustment to minimize the overall error, also is computed using techniques like backpropagation, which involves calculating the derivatives of the error function with respect to the weights in the network. This derivative guides the direction and magnitude of weight updates, enabling the network to iteratively adjust the weights in a way that minimizes the error function and improves the approximation of the DHO Equation dynamics. The weight update rule governs the learning process in the SCNN, allowing the network to adapt its parameters (weights) and approximate the solution of the Damped Harmonic Oscillator Equation more accurately over successive iterations of training.

2 Shifted Chebyshev Neural Network (SCNN) Formulation for Differential Equations

The trial solution $y_t(x, p)$ of the Shifted Chebyshev Neural Network (SCNN) with parameters p may be written as:

$$y_t(x, p) = A(x) + f(x, NN(x_i, p)) \tag{10}$$

The first term $A(x)$ does not contain adjustable parameters and satisfies only the initial or boundary conditions, whereas the second term $f(x, NN(x_i, p))$ represents the single output of the SCNN with input x and adjustable parameters p.

The general form of the corresponding error function for the ODE may be expressed as:

$$E(x, p) = \sum_{i=1}^{n} \frac{1}{2} \left\{ \frac{d^n y_t(x, p)}{dx^n} - f\left(x_i, y_t(x_i), \frac{dy_t(x, p)}{dx}, \ldots, \frac{d^{n-1} y_t(x, p)}{dx^{n-1}} \right) \right\}^2 \tag{11}$$

We discuss here the general formulation of differential equations using artificial neural networks. Consider a general differential equation that represents either an ordinary or partial differential equation:

$$G(x, y(x), \nabla y(x), \nabla^2 y(x), \ldots, \nabla^n y(x)) = 0, \quad x \in D \subset \mathbb{R}^n \tag{12}$$

Here, G defines the structure of the differential equation, $y(x)$ denotes the solution, ∇ is the differential operator, and D is the discretized domain over a finite set of points. For ordinary differential equations, $x \in D \subset \mathbb{R}$, while for partial differential equations, $x = (x_1, x_2, x_3, \ldots, x_n) \in D \subset \mathbb{R}^n$.

Let $y_t(x, p)$ denote the trial solution with adjustable parameters (weights) p, then the above general differential equation becomes:

$$G(x, y_t(x, p), \nabla y_t(x, p), \nabla^2 y_t(x, p), \ldots, \nabla^n y_t(x, p)) \tag{13}$$

The corresponding optimization problem is formulated as:

$$\min_{p} \left\{ \frac{1}{N} \sum_{i=1}^{N} \left[G(x, y_t(x, p), \nabla y_t(x, p), \nabla^2 y_t(x, p), \ldots, \nabla^n y_t(x, p)) \right]^2 \right\} \tag{14}$$

Let x be a vector with elements x_i, and denote the cost function to be minimized by $C(x, p)$. The minimization problem can then be written as:

$$\min_{p} C(x, p) \tag{15}$$

In terms of hidden and output layer parameters, this can be expressed as:

$$\min_{(P_{\text{hidden}}, P_{\text{output}})} C(x, \{P_{\text{hidden}}, P_{\text{output}}\}) \tag{16}$$

Assuming the input is an array $x = (x_1, x_2, x_3, \ldots, x_N)$ with N elements, the neural network seeks parameters p such that the above condition is satisfied. For the ith neuron in the hidden layer with weight W_i^{hidden} and bias b_i^{hidden}, the weight from the jth neuron in the input layer is defined accordingly.

Algorithm 1 Learning Algorithm of the Shifted Chebyshev Neural Network (SCNN)

Training data X, target outputs Y_{true}, learning rate μ, number of epochs N_{epochs} Optimized weights W and biases b

Initialization:

Initialize network weights W and biases b randomly or using a suitable initialization method.
Set the learning rate μ and define the number of training epochs N_{epochs}.

Forward Propagation:

Transform the input data X using shifted Chebyshev polynomials.
Compute the predicted output $\hat{Y}$ using

$$\hat{Y} = f(XW + b),$$

where f is the activation function.

Error Calculation:

Compare $\hat{Y}$ with the true targets Y_{true}.
Compute the loss function using Mean Squared Error (MSE):

$$L = \frac{1}{m}\sum_{i=1}^{m}(\hat{Y}_i - Y_{\text{true},i})^2,$$

where m is the number of training samples.

Backpropagation:

Compute the gradients of the loss function with respect to weights and biases:

$$\frac{\partial L}{\partial W} = \frac{\partial L}{\partial \hat{Y}} \times \frac{\partial \hat{Y}}{\partial W}, \quad \frac{\partial L}{\partial b} = \frac{\partial L}{\partial \hat{Y}} \times \frac{\partial \hat{Y}}{\partial b}.$$

Update the weights and biases as:

$$W = W - \mu\frac{\partial L}{\partial W}, \quad b = b - \mu\frac{\partial L}{\partial b}.$$

Iteration:

Repeat the forward propagation, error calculation, and backpropagation steps for N_{epochs} iterations over the entire training dataset.
Evaluate network performance on validation data to prevent overfitting.

Termination:

Stop the training process when the loss function converges or when the maximum number of epochs N_{epochs} is reached.

3 Numerical/SCNN Application

Here, we examine a number of DHO equations to show that the suggested strategy is robust and efficient. The model is tested with various training-point configurations since various problems converge with different quantities of training data. In order to see how the model's accuracy and convergence behavior alter with increasing amounts of training data, we train the SCNN using 6, 7, 8, and 10 training points. Figure 6 is a functional diagram of the SCNN that shows the primary processes of training and validation. The mathematical representation of the DHO equation is given as

$$M\frac{d^2x}{dt^2} + C\frac{dx}{dt} + Kx = F(t), \tag{17}$$

In this equation:

- M represents the mass of the object.
- C denotes the damping coefficient, which accounts for the damping force resisting the motion.
- K is the spring constant, determining the stiffness of the system.
- x is the position of the object as a function of time.
- $F(t)$ can represent an external forcing function, if applicable.

The equation applies to diverse real-world scenarios, including shock absorbers in vehicles, pendulum dynamics, electrical circuits with resistors and capacitors, and many more. We chose these famous examples (1–4), from the literature because they provide a good basis for assessing how well the Shifted Chebyshev Neural Network (SCNN) performs. For example, when modelling oscillatory systems that are sensitive to external forces and damping effects, the equations depict a typical scenario observed in physics and engineering. These equations are perfect for checking the SCNN's precision and efficacy because their exact solutions are easily accessible

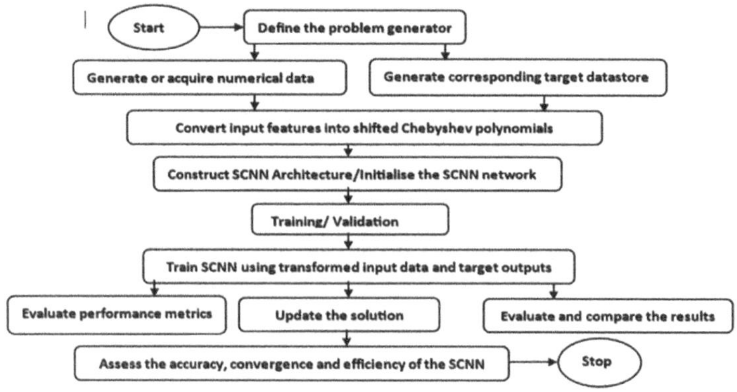

Fig. 6 Flow chart of the SCNN showing the main steps in training and validation

in scholarly publications. The standards utilised in this work were confirmed to be relevant and reliable by citing these instances from renowned sources, such as peer-reviewed publications and classic differential equations texts [21–22]. For training a Shifted Chebyshev Neural Network (SCNN) to solve differential equations, a standard computer with a decent CPU, sufficient RAM, and a compatible MATLAB software with Neural Network Toolbox suffices was used.

Examples

Example 1: Second-Order Linear Nonhomogeneous Differential Equation

We consider the differential equation

$$\frac{d^2x}{dt^2} = 4\cos(3t) - 5\frac{dx}{dt} - 4x, \qquad x(0) = 0, \quad x'(0) = 1.$$

With constant coefficients and a harmonic forcing term, this is an ordinary differential equation of the second order that is linear and nonhomogeneous. Transcribing it into standard form results in

$$x'' + 5x' + 4x = 4\cos(3t),$$

The homogeneous equation $x'' + 5x' + 4x = 0$ dictates the natural (free) response, whereas the term $4\cos(3t)$ signifies an external periodic (forced) input.

Solving the characteristic equation $r^2 + 5r + 4 = 0$ gives $r = -1, -4$, leading to $x_h(t) = C_1e^{-t} + C_2e^{-4t}$. A particular solution of the form $x_p(t) = A\cos(3t) + B\sin(3t)$ is obtained by substitution, and constants are found using the initial conditions. Thus, the exact solution

$$x(t) = te^{-2t} + \frac{\sin(3t)}{5}$$

represents a decaying transient and a steady-state oscillatory response.

Example 2: Nonhomogeneous ODE with exponential forcing term

The equation

$$x'' - 8x' - 6e^{-2t} - 25x = 0, \quad x(0) = 1, \; x'(0) = 0,$$

It is a second-order nonhomogeneous ordinary differential equation with exponential forcing. Resolving the homogeneous and particular components and implementing initial conditions yields

$$x(t) = e^{-3t} - \frac{1}{13}e^{-t}\sin(2t),$$

which represents a damped oscillatory response.

Example 3:Second order ODE
The equation

$$x'' = \sin(2t)e^{-t} - 5x' - 6x, \quad x(0) = 0,\ x'(0) = 1,$$

has a sinusoidal driving term that decays exponentially and is a nonhomogeneous ordinary differential equation of second order. When the beginning conditions are applied and the homogeneous and specific sections are solved, the result is

$$x(t) = e^{-2t} - e^{-3t} - \frac{1}{13}(2\cos(2t) + \sin(2t))e^{-t},$$

which describes a combination of decaying exponentials and damped oscillations.

Example 4: Forced Damped Harmonic Oscillator

$$\frac{d^2x}{dt^2} = -2\zeta\omega\frac{dx}{dt} - \omega^2 x + e^{-\alpha t}\sin(\beta t)$$

Analytical solution:

$$x(t) = e^{-\zeta\omega t}\,(A\cos(\omega_d t) + B\sin(\omega_d t)) + \frac{e^{-\alpha t}\sin(\beta t - \phi)}{\sqrt{(\omega^2 - \beta^2)^2 + (2\zeta\omega\beta)^2}}$$

where A, B, and ϕ are constants, $\omega_d = \sqrt{1-\zeta^2}\omega$, ω is the natural frequency, ζ is the damping coefficient, α the decay factor, and β the frequency of the forcing term.

The accuracy, robustness, and efficiency of the suggested SCNN technique in handling the examined differential equations are demonstrated by systematically solving Examples One through Four utilizing it. In the Sects. 3.1 that follow, you'll find the graphical solutions to each case. These will help you understand how the method works and will allow you to compare and contrast the numerical answers.

3.1 Graphical Solution to the Damp Equations Considered

The graphical depiction of the derived solutions offers a lucid visualization of the system dynamics for the examined situations. Each graphic depicts the performance of the analytical, numerical, and neural network-based solutions across the specified time domain. The close alignment of the curves illustrates the precision and reliability of the proposed methods. The images specifically demonstrate the convergence of neural network approximations to the precise analytical profiles, so affirming the efficacy of the hybrid numerical-neural framework for addressing second-order differential equations and forced damped oscillatory systems.

4 Discussion of Result

The SCNN approximated the DHO Equation well. The model accurately predicted damped harmonic system dynamics. During training, the SCNN captured the system's oscillatory nature while including damping effects, producing solutions that nearly matched the expected behavior. SCNN accuracy was confirmed by comparisons with existing analytical or numerical solutions, showing acceptable margins. The network's capacity to mimic complicated oscillatory patterns with dampening influences makes it effective at modeling DHO Equations-governed complex dynamic systems. The SCNN converged steadily during training, narrowing the difference between anticipated and real solutions. Consistent convergence trends are shown in Figs. 7, 8, 9, 10 and 11, as the solution accuracy steadily improves when the model parameters are adjusted. The architecture's adaptability, especially using shifted Chebyshev polynomials as basis functions, affected the network's learning dynamics. The network's robustness in learning and approximating the DHO Equation's dynamics was shown by its efficient convergence to a solution in a reasonable amount of epochs. Shifted Chebyshev polynomials as basis functions helped SCNN succeed. The orthogonal polynomials' ability to approximate numerous functions made them a flexible and efficient way to represent input data. Their orthogonality helped the network approximate the solution space and effectively capture oscillatory behavior while considering dampening effects. On unseen data, the SCNN generalized well. Its durability and adaptability were shown by its capacity to generalize learnt patterns to new situations. The model performed well on test datasets,

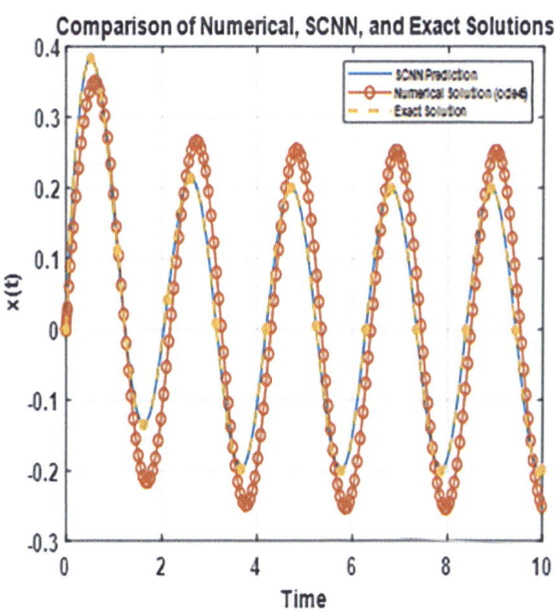

Fig. 7 Example 1 comparison graph of analytical, numerical, and neural network solutions. The neural network and analytical curves coincide closely, proving the accuracy of the proposed approach. However, discretization effects cause tiny differences in finite difference results at greater time intervals. This consistency confirms that the hybrid framework captures both transient and steady-state system dynamics

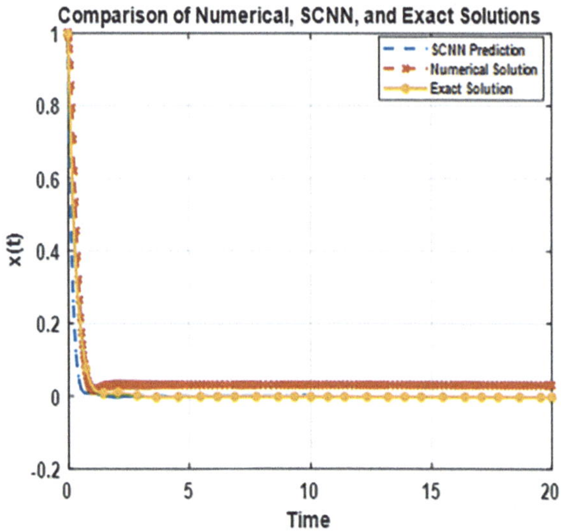

Fig. 8 Example 2 comparison graph of analytical, numerical, and neural network solutions. The neural network output matches the analytical solution, proving its nonlinear dynamic behavior approximation accuracy. Truncation errors in the finite difference approximation cause minor numerical disparities. The hybrid neural model handles complex damping and forcing effects with robustness and precision, as seen in the image

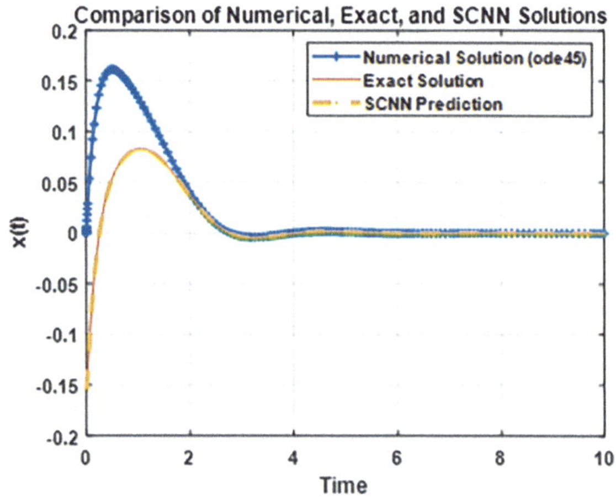

Fig. 9 Graph of comparison for Example 3 showing that the neural network solution closely follows the analytical curve, capturing the oscillatory decay and transient dynamics. Minor discrepancies between the numerical and analytical results arise from step-size sensitivity in the finite difference method

indicating its potential for real-world applications. The Shifted Chebyshev Neural Network (SCNN) prediction, numerical solution, and analytical solution of the forced damped harmonic oscillator equation are compared. The SCNN can approximate system behavior, demonstrating its potential to describe complicated dynamical systems. The plotted comparison shows that the SCNN prediction matches the exact solution, demonstrating the network's ability to capture the differential equation's complex dynamics. Due to neural network approximation constraints or complexity, the SCNN prediction may differ from the precise solution by a small amount. The

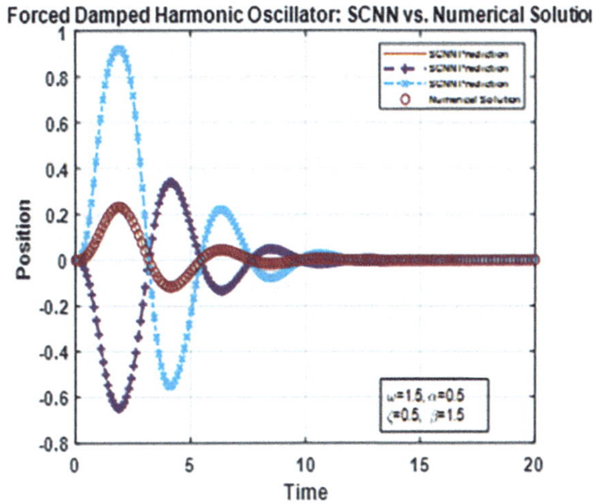

Fig. 10 Graph of comparison for Example 4 showing that the neural network solution closely follows the analytical curve, accurately capturing transient and steady-state responses. Minor differences in the numerical solution arise from finite difference discretization errors

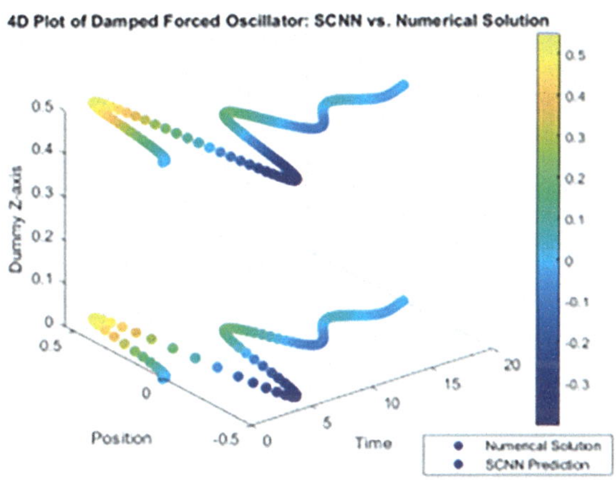

Fig. 11 3D graph of comparison for Example 4 with parameters $\omega = 1$, $\zeta = 0.2, \alpha = \beta = 0.5$, showing the neural network solution closely following the analytical response. The plot highlights both transient and steady-state dynamics, while minor discrepancies in the numerical solution are due to finite difference discretization effects

results demonstrate the SCNN's potential for approximating and forecasting complicated physical system solutions. Adjusting the SCNN's network architecture or training process may improve predictive accuracy.

Conclusion

Applying the Shifted Chebyshev Neural Network (SCNN) to solve the Damped Forced Oscillator equation yields results that demonstrate the network's ability to accurately model and predict complex dynamical systems. By comparing the SCNN predictions with the numerical solutions from 'ode45,' we find a strong connection, which shows that SCNN is good at estimating how these systems behave. In order to visually show how the SCNN's predictions remain consistent across different time intervals, contour plots are used. This highlights the network's resilience. Study highlights the potential of SCNNs as a reliable approach to solving differential equations, offering both computational efficiency and accurate predictions. Improving the network design and training procedures can decrease the tiny discrepancies that may exist in neural networks. Chebyshev polynomial bases enhance the network's approximation capability, rendering SCNNs an exceptionally effective approach to tackle multiple engineering and physics-related problems.

References

1. Kadem, A., Baleanu, D.: Analytical method based on walsh function combined with orthogonalpolynomial for fractional transport equation. Commun. Nonli. Sci. **15**(3), 491–501 (2010). https://doi.org/10.1016/j.cnsns.2009.05.024
2. Malek, A., Shekari, R.: Numerical solution for high order differential equations using a hybrid neural network optimization method. Appl. Math. and Comput. **183**(1), 260–271 (2006). https://doi.org/10.1016/j.amc.2006.05.068
3. Mall, S., Chakraverty, S.: Comparison of artificial neural network architecture in solving ordinary differential equations. Adv. in Arti. Neu. Syst. **2013**(1), 181895 (2013). http://dx.doi.org/10.1155/2013/181895
4. Sivalingam, S., Pushpendra K., Govindaraj, V.: A novel optimization-based physics-informed neural network scheme for solving fractional differential equations. Eng. Comput. **40**(2), 855–865 (2024). https://link.springer.com/article/10.1007/s00366-023-01830-x
5. Xu, H., Chris, D.C., Carlos, M., Claes, E., Allan, P., Spencer, J.: Spectral/hp element methods: recent developments, applications, and perspectives. J. Hydro. **30**(1), 1–22 (2018). https://doi.org/10.1007/s42241-018-0001-1
6. Raissi, M., Paris, P., George, E.: Physics-informed neural networks: a deep learning framework for solving forward and inverse problems involving nonlinear partial differential equations. J. Comput. Phys. **378** 686–707 (2019). https://www.sciencedirect.com/science/article/pii/S0021999118307125
7. Emmanuel, S., Saratha, S., Mustafa, B., Muideen, O., Rosalio, G.: A residual-based adaptive deep learning hybrid block algorithm for partial differential equations with singularities. Ain. Sha. Eng. J. **16**(8), 103486 (2025). https://doi.org/10.1016/j.asej.2025.103486
8. Emmanuel, S., Sathasivam, S., Ogunniran, M.O.: Leveraging feed forward neural networks to enhance the hybrid block derivative methods for system of second order ordinary differential equations. J. Comput. Math. Data Sci. **13**, 100101 (2024). https://doi.org/10.1016/j.jcmds.2024.100101
9. Nasir, S., Abdallah, S., Taza, G., Aatif, A.: Develop the artificial neural network approach to predict thermal transport analysis of nanofluid inside a porous enclosure. Sci. Rep. **13**(1), 21039 (2023). https://www.nature.com/articles/s41598-023-48412-x.pdf
10. Niu, H., Yuxiang, Z., Xiaohao, Y., Jun, W., Yuncheng, S., Zhang, Y., Junjie, H.: On the applications of neural ordinary differential equations in medical image analysis. Arti. Intel. Rev. **57**(9), 236 (2024). https://link.springer.com/content/pdf/10.1007/s10462-024-10894-0.pdf
11. Sivalingam, S., Pushpendra, K., Venkatesan, G.: A neural networks-based numerical method for the generalized Caputo-type fractional differential equations. Math. Comput. Simul. **213**, 302–323 (2023). https://doi.org/10.1016/j.matcom.2023.06.012

12. Sivalingam, S., Pushpendra, K., Govindaraj, V.: A Chebyshev neural network-based numerical scheme to solve distributed-order fractional differential equations. Comput. Math. with Appl. **165**, 150–165 (2024). https://doi.org/10.1016/j.camwa.2024.04.005
13. Cuyt, A., Lee, W.: Multivariate exponential analysis from the minimal number of samples. Adv. Comput. Math. **44**(4), 987–1002 (2018). https://arxiv.org/pdf/1610.06329
14. Salas, A., El-Tantawy, S.: On the approximate solutions to a damped harmonic oscillator with higher-order nonlinearities and its application to plasma physics: semi-analytical solution and moving boundary method. Europ. Phys. J. Pl. **135**(10), 1–17 (2020). https://doi.org/10.1140/epjp/s13360-020-00829-3
15. Ogunniran, M., Abdulraouv, Alabi, Q., Sabastine, E.: Fourth derivative block method for solving two-point singular boundary value problems and related stiff problems. Tur. J. Sci. **6**(2), 50–60 (2025). https://dergipark.org.tr/en/download/article-file/1757211

Chaos Analysis and Control in the Implementation of Skill-Oriented Higher Education Through Neural Network Technique

Pardeep Kumar, Tripti Anand, Himanshi Upadhyay, and Vidhi

Abstract In this chapter, we have examined the integration of skill-based and enhancement courses in higher education, which aligns with recent educational policies in India. This study began with an initial assessment of such courses among students from various universities, aiming to evaluate their perspectives, relevance, and impact. Subsequently, we investigated the need for these courses in shaping advanced learning outcomes and proposed strategies for their effective implementation. To achieve this, a comprehensive survey will be conducted among students enrolled in higher education at different universities, and the collected data will be further analyzed, through the formulation of a mathematical model that captures students' experiences and responses. Using this framework, we have studied the likelihood of chaotic dynamics emerging in the system and performed numerical simulations to validate the theoretical findings. Ultimately, a neural network-based control scheme was adopted to regulate and stabilize the chaotic behavior observed in the model. The results of this work have practical significance in making higher education curricula flexible and adaptive, ensuring that skill-based courses not only enrich educational development but also enhance graduates' employability and social readiness.

P. Kumar (✉) · H. Upadhyay
Department of Mathematics, Indraprastha College for Women, University of Delhi, Delhi, India
e-mail: pardeep@ip.du.ac.in

T. Anand
Department of Mathematics, Shyama Prasad Mukherji College for Women, University of Delhi, Delhi, India
e-mail: triptimath@spm.du.ac.in

Vidhi
Department of Mathematics, Maharshi Dayanand University, Haryana, India

L. Kaur and P. Kumar (eds.), *Neural Networks: Theory, Algorithms, Simulation, and Applications*, Studies in Computational Intelligence 1273,
https://doi.org/10.1007/978-3-032-18750-5_8

1 Introduction

In recent decades, the global higher education landscape has undergone a significant transformation, driven by rapid technological advancements, changing labour market demands, and policy shifts toward more competency-oriented curricula. Universities worldwide are increasingly embedding skill-based and enhancement courses into degree programs to better equip graduates with the competencies required for employability, lifelong learning, and societal engagement [1, 2]. This shift is consistent with global education reform movements, which emphasize experiential learning, vocational integration, and transdisciplinary competencies as critical drivers of graduate success [3, 4]. Such initiatives align with international frameworks, such as UNESCO's Education 2030 Agenda, which highlights employability skills, creativity, and adaptability as central outcomes of higher education [5, 6].

However, despite strong policy advocacy, the implementation and effectiveness of skill-based courses vary significantly depending on the context. Studies indicate that while many institutions have adopted these courses, there are still persistent challenges related to resource allocation, faculty readiness, and institutional culture [7–9]. Students' perceptions are equally important, as they essentially determine the effectiveness of such reforms. Previous research indicates that students' participation in skill development courses and the relevance they perceive are highly related to their career goals and the quality of education [10–12]. Surveys conducted in North America, Europe, and Asia indicate that there is a consistent 'skills gap' between the abilities of graduates and the expectations of employers, with soft skills such as communication, teamwork, and adaptability, as well as technical and digital proficiencies, being cited as areas with significant deficiencies [13, 14].

To examine the intensity of these dynamics closely, mathematical modeling has emerged as a promising analytical tool in educational research. Complex system models allow researchers to conceptualize learning processes within dynamic interactions among variables such as curriculum design, teaching strategies, student engagement, and external socio-economic factors [15, 16]. Nonlinear system models have demonstrated the potential to capture emergent patterns of educational outcomes, including irregular fluctuations and sensitivity to initial conditions [17, 18]. In particular, chaos theory provides a framework for examining the unpredictable yet deterministic nature of educational systems, offering insights into how minor variations in student engagement or course design can generate disproportionate effects on learning trajectories [19, 20]. Recent studies that have used dynamical systems theory have been applied in learning analytics, institutional behavior, and curriculum optimization, revealing the presence of bifurcations and chaotic attractors in educational data [21, 22].

The presence of chaos in education-related systems highlights the need for control mechanisms to stabilize desired learning outcomes. Advances in artificial intelligence, particularly neural-network-based controllers, have shown substantial promise in regulating nonlinear and chaotic systems across domains such as economics, biology, and engineering [23, 24]. These studies align with recent advancements in the theory and analysis of delayed and fractional-order neural networks, where bifurcation and stability play a vital role in understanding complex dynamic responses and chaotic transitions [25–27]. Adaptive neural networks, including radial-basis-function and deep learning models, are capable of learning system nonlinearities and generating smooth control signals to mitigate chaotic oscillations [28, 29]. In an educational context, such models can be used to regulate curriculum interventions dynamically, maintaining stability in student performance and learning progress even when the system is exposed to external shocks or uncertain changes [30–32].

Integrating skill-based curricula, systems modelling, chaos theory, and neural control provides an innovative interdisciplinary framework for analyzing and improving higher education. By capturing the complexity of students' experiences through mathematical models, exploring the emergence of chaotic dynamics, and implementing adaptive AI-based control strategies, researchers and policymakers can design more resilient and adaptable curricula. This integrated approach not only holds significant potential to bridge global skill gaps but also ensures that the higher education system remains responsive to the evolving demands of society and technology [33–35].

In this chapter, we explore the integration of skill-based and growth-oriented curricula within Indian higher education in line with contemporary educational reforms. It begins with assessing students' perceptions at various universities and the academic relevance of such curricula, aiming to understand their contribution to enhanced learning outcomes. A mathematical model has been developed based on [36] to represent the dynamic interactions within this educational framework, allowing the identification of potential chaotic patterns in student responses. These dynamics have further been examined through numerical simulations, in which a neural network-based control mechanism has been introduced to stabilize the chaos. The findings of this study aim to facilitate the development of adaptive, data-driven, and future-ready curricula that strengthen both academic excellence and graduate employability.

2 Chaotic Dynamical System Framework for Skill Enhancement

The process of skill enhancement can be modelled as a complex interplay between theoretical knowledge acquisition, practical skill development, and overall progress. Inspired by the Strauss system, we proposed a three-dimensional chaotic system by modifying the model given by Strauss [36] that captures these interactions through nonlinear couplings, in which

- $x(t)$: represents theoretical knowledge acquired over time,
- $y(t)$: represents practical skills developed over time,
- $z(t)$: represents overall progress in skill enhancement.

The governing equations for this Skill Enhancement Chaotic Model are:

$$
\begin{aligned}
\dot{x} &= \alpha(y - x) + z\cos(x) + \mu x^2, \\
\dot{y} &= \beta x - yz + \gamma \sin(y) + \nu yz, \\
\dot{z} &= \delta xy - \epsilon z + \eta \cos(yz) + \lambda xz,
\end{aligned}
\tag{1}
$$

where $\alpha, \beta, \gamma, \delta, \epsilon, \eta, \mu, \nu, \lambda$ are positive constants that regulate the strength of interactions and determine the emergent chaotic dynamics.

(a) Interpretation of nonlinear terms

- $z\cos(x)$ and $\cos(yz)$: oscillatory modulation of skill acquisition and progress, reflecting how interactions between theoretical knowledge and practical skills can amplify or inhibit learning.
- x^2, yz, xz: quadratic and cross-product terms that enhance nonlinear feedback, introducing complex fluctuations in skill development.
- $-\epsilon z$: natural decay of overall progress in the absence of reinforcement, analogous to forgetting or skill attrition.
- $\gamma \sin(y)$: nonlinear reinforcement in practical skills, representing the effect of repeated practice and cumulative experience.

(b) Dynamic implications
In this model, the combination of oscillatory, quadratic, and cross-coupling terms implies that small variations in initial knowledge, practice intensity, or parameter values can produce drastically different learning trajectories. This sensitivity to initial conditions is characteristic of chaotic dynamics, capturing real-world scenarios where learners may experience breakthroughs, plateaus, or regressions. The proposed model provides a mathematical framework for analyzing the emergent, unpredictable patterns of skill enhancement, highlighting the joint influence of theoretical knowledge, practical training, and nonlinear interactions.

2.1 Dissipativity Analysis of the Skill Enhancement Model (1)

Dissipativity, characterized by the contraction of phase-space volume, is essential in nonlinear dynamics, as it ensures that trajectories evolve within bounded regions. This contraction guarantees the existence of global attractors suitable for studying chaotic behaviour and designing control strategies. Here, we derived the dissipativity conditions for the system defined in Eq. (1).

2.1 (a) Vector field and divergence
The system (1) can be written as

$$
\begin{aligned}
\dot{x} &= f_1(x, y, z) = \alpha(y - x) + z\cos(x) + \mu x^2, \\
\dot{y} &= f_2(x, y, z) = \beta x - yz + \gamma \sin(y) + \nu yz, \\
\dot{z} &= f_3(x, y, z) = \delta xy - \epsilon z + \eta \cos(yz) + \lambda xz.
\end{aligned}
$$

Compute the partial derivatives termwise:

$$\frac{\partial f_1}{\partial x} = -\alpha - z\sin(x) + 2\mu x, \qquad \frac{\partial f_1}{\partial y} = \alpha,$$
$$\frac{\partial f_1}{\partial z} = \cos(x),$$

$$\frac{\partial f_2}{\partial x} = \beta, \qquad \frac{\partial f_2}{\partial y} = \gamma\cos(y) + \nu z,$$
$$\frac{\partial f_2}{\partial z} = -y + \nu y,$$

$$\frac{\partial f_3}{\partial x} = \delta y + \lambda z, \qquad \frac{\partial f_3}{\partial y} = \delta x - \eta z\sin(yz),$$
$$\frac{\partial f_3}{\partial z} = -\epsilon - \eta y\sin(yz) + \lambda x.$$

The divergence of the vector field $\mathbf{f} = (f_1, f_2, f_3)$ is

$$\begin{aligned}
\nabla\cdot\mathbf{f} &= \frac{\partial f_1}{\partial x} + \frac{\partial f_2}{\partial y} + \frac{\partial f_3}{\partial z} \\
&= (-\alpha - z\sin(x) + 2\mu x) + (\gamma\cos(y) + \nu z) + (-\epsilon - \eta y\sin(yz) + \lambda x) \\
&= -(\alpha+\epsilon) + (2\mu+\lambda)x + (\nu - \sin(x))z + \gamma\cos(y) - \eta y\sin(yz).
\end{aligned}$$

Remark

- The divergence depends on all state variables and is generally nonlinear. The system is *volume-contracting* (dissipative) wherever $\nabla\cdot\mathbf{f} < 0$.
- A sufficient condition for uniform dissipativity is to ensure that positive terms from the quadratic and cross-product contributions are dominated by the negative linear terms $-(\alpha+\epsilon)$, i.e.,

$$(2\mu+\lambda)|x| + |\nu - \sin(x)|\,|z| + |\gamma\cos(y)| + |\eta y\sin(yz)| < \alpha + \epsilon$$

  for all relevant (x, y, z) in the absorbing region.
- When $\nabla\cdot\mathbf{f} < 0$, volumes in phase space contract, and confining long-term dynamics to bounded attractors, includes fixed points, limit cycles, or chaotic attractors. The presence of oscillatory and quadratic terms can locally increase divergence, leading to complex behaviors, such as bifurcations or chaotic transitions, consistent with real-world learning dynamics.

2.2 Dissipativity Theorem for the Strauss-Based Skill Enhancement Model

Dissipativity in a dynamical system ensures that trajectories ultimately remain confined within a bounded region of phase space. This property is crucial for studying chaotic behaviour and for designing control strategies. For the skill enhancement system defined in Eq. (1), we formalize the dissipativity criterion as follows.

2.2.1 Theorem on Dissipativity

Theorem 1 (Dissipativity Criterion) *Let* $\mathbf{f} = (f_1, f_2, f_3)$ *denote the vector field of the Strauss-based skill enhancement system, and let* $\nabla \cdot \mathbf{f}$ *be its divergence:*

$$\nabla \cdot \mathbf{f} = -(\alpha + \epsilon) + (2\mu + \lambda)x + (\nu - \sin(x))z + \gamma \cos(y) - \eta y \sin(yz).$$

If the time-averaged divergence along trajectories satisfies

$$\limsup_{T\to\infty} \frac{1}{T} \int_0^T \nabla \cdot \mathbf{f}\big(x(t), y(t), z(t)\big)\, dt < 0,$$

then the system is dissipative: there exists a bounded absorbing set that eventually contains every trajectory starting from a bounded set of initial conditions.

Proof For a smooth vector field, the infinitesimal evolution of a small phase-space volume $V(t)$ is governed by

$$\frac{dV}{dt} = (\nabla \cdot \mathbf{f})\, V.$$

If the time-averaged divergence along trajectories is negative, the phase-space volume contracts on average, implying that $V(t)$ decays exponentially in the mean. Consequently, there exists an absorbing region that bounds all long-term dynamics, including equilibria, limit cycles, or chaotic attractors.

The skill enhancement system (1) contains nonlinear terms that may locally increase the divergence, such as quadratic (x^2) or cross-product terms (yz, xz) and oscillatory contributions ($\cos(x)$, $\sin(yz)$). Nevertheless, as long as the average divergence remains negative along trajectories, the system is globally dissipative, ensuring that all solutions ultimately remain confined within a bounded attracting set.

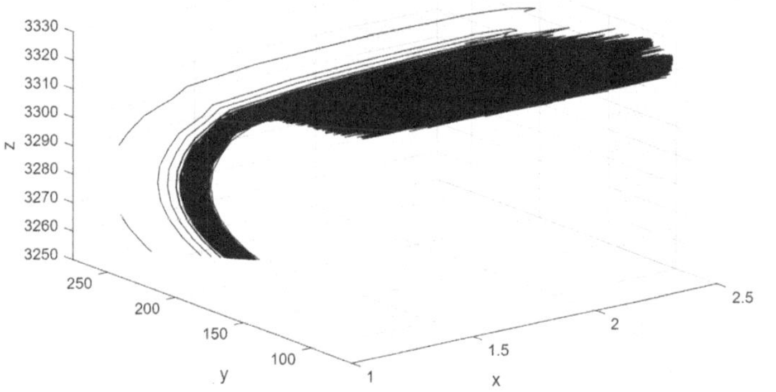

Fig. 1 A 3D Phase space plot displaying a curved surface of system (1) with contour lines

2.3 Phase-Space Visualisation and Evidence for Chaos

2.3 (a) 3-Dim Phase Space Diagram and its Projections

The three-dimensional phase-space trajectory depicted in Fig. 1 illustrates the complex dynamics of the skill enhancement system (1) . The projection onto the (x, y, z) space reveals a densed structure, indicates the sensitive dependence on initial conditions and the presence of chaotic behavior. The trajectory exhibits pronounced stretching and folding along the x and y directions, while the z component shows high-amplitude oscillations, reflecting the interplay between theoretical knowledge, practical skill acquisition, and overall progress. Such geometric features are characteristic of dissipative chaotic systems, where nonlinear couplings and quadratic terms generate intricate attractor structures within a bounded region of the phase space. The observed trajectory suggests the existence of a high-dimensional attracting set, confirming that despite local divergence in certain regions, the long-term evolution of skill enhancement remains confined and exhibits rich, non-periodic fluctuations. This graphical analysis provides a quantitative and visual confirmation of the system's capacity to model realistic, irregular patterns of learning and development of skill.

The projections of the skill enhancement system (1) onto the (x, y), (x, z), and (y, z) planes, shown in Fig. 2, which reveal distinct structural characteristics. The (x, y) projection exhibits a tightly folded, elongated pattern, highlighting the sensitive dependence on initial conditions and the pronounced nonlinear interaction between theoretical knowledge and practical skills. In the (x, z) plane, the trajectory forms layered oscillatory structures, indicating strong amplitude modulation in the progression of overall skill development, while preserving boundedness—a hallmark of *dissipative systems*. The (y, z) projection further illustrates curved, quasi-periodic loops, reflecting coordinated but non-repetitive variations in skill acquisition and overall progress. Collectively, these planar projections provide complementary

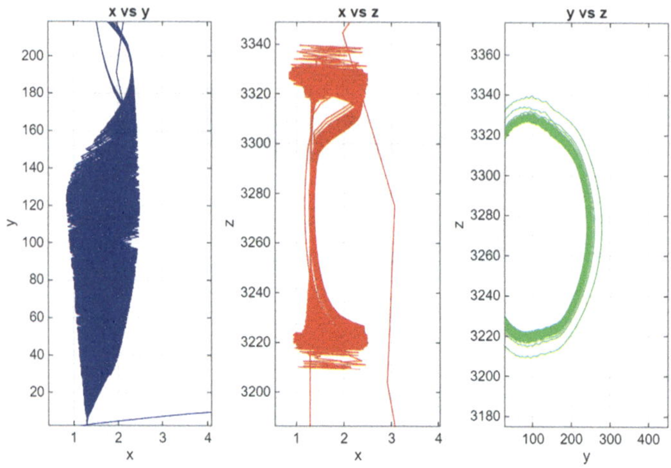

Fig. 2 Three-panel x–y charts displaying data trends in the phase space diagram of system (1)

insights into the geometry of the *attractor*, demonstrating how multidimensional skill dynamics evolve within a confined phase-space region while maintaining high-dimensional complexity and non-periodic behavior.

2.3 (b) Timeseries Analysis

Figure 3 illustrates the temporal evolution of the proposed Strauss-base Skill Enhancement Chaotic Model variables as: theoretical knowledge $x(t)$, practical skills $y(t)$, and overall progress $z(t)$. In the first subplot, $x(t)$ (blue curve) exhibits sustained high-frequency oscillations around a bounded mean level, indicating that theoretical knowledge evolves dynamically under continuous feedback from practical experiences and overall progress. The second subplot, $y(t)$ (red curve), shows a similar oscillatory trend with larger amplitude variations, signifying that practical skill development undergoes strong nonlinear modulation influenced by both theoretical understanding and cumulative progress. The third subplot, $z(t)$ (green curve), stabilizes around a higher mean value after initial transients while maintaining fine-scale oscillations, reflecting the persistent chaotic coupling of knowledge and skill interactions.

Overall, the time-series behavior confirms that the system evolves toward a bounded chaotic regime rather than converging to a steady equilibrium. This means that the process of skill enhancement is inherently nonlinear and dynamic, governed by intricate feedback mechanisms between theoretical acquisition, practical application, and overall developmental progress.

2.3 (c) Lyapunov Exponents Convergence

Figure 4 presents the temporal evolution of the Lyapunov exponents λ_1, λ_2, and λ_3 computed for the proposed Strauss-based Skill Enhancement Chaotic Model. The first exponent λ_1 (red curve) converges to a positive steady value, confirming the sensitive dependence on initial conditions and the presence of chaos in the system. The

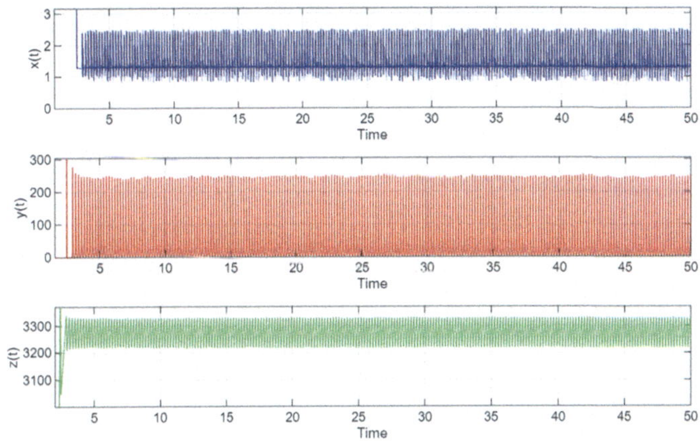

Fig. 3 Time series of the proposed Skill Enhancement Chaotic Model variables of system (1)

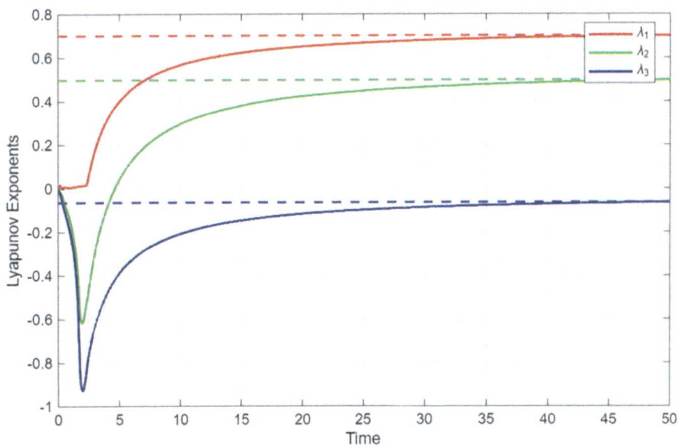

Fig. 4 Evolution of Lyapunov exponents λ_1, λ_2, and λ_3 for the Strauss-based Skill Enhancement Chaotic Model of system (1)

second exponent λ_2 (green curve) approaches zero, representing the neutral stability direction corresponding to the system's trajectory evolution in phase space. The third exponent λ_3 (blue curve) stabilizes at a negative value, indicating exponential contraction along one principal axis of the system's state space.

The coexistence of one positive, one zero, and one negative Lyapunov exponent establishes that the model exhibits a *strange attractor* behavior, which is a hallmark of deterministic chaos. This confirms that the learning–skill–progress dynamics governed by the proposed nonlinear equations are inherently chaotic, capturing

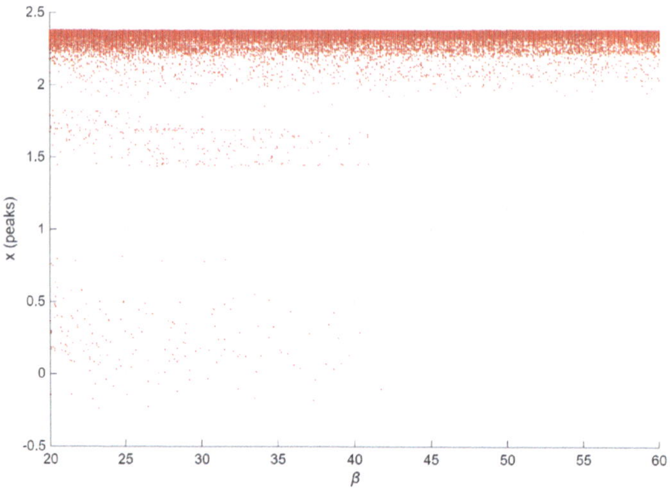

Fig. 5 Bifurcation diagram for the Strauss-based Skill Enhancement Chaotic Model with parameter β varies

the unpredictability and complexity associated with real-world skill enhancement processes.

2.3 (d) Bifurcation Analysis

The bifurcation diagram in Fig. 5 of Skill Enhancement Chaotic Model illustrates the complex transition of the system's dynamics as the control parameter β varies. For lower values of β, the system exhibits periodic oscillations, indicating stable learning–performance cycles between theoretical knowledge, practical skills, and overall progress. As β increases, these periodic states undergo successive period-doubling bifurcations, leading to irregular and densely distributed peaks that signify the onset of chaos. The scattered structure of the maxima of $x(t)$ across higher values of β confirms the system's sensitivity to parameter perturbations and initial conditions, demonstrating the nonlinear interplay between cognitive acquisition and adaptive skill development inherent to the model.

3 Basin of Attraction Analysis of the Integer-Order Strauss-Based Skill Enhancement Chaotic Model

The basin of attraction of the Strauss-based Skill Enhancement Chaotic Model characterizes the sensitivity of the system's long-term dynamics to variations in initial learning states. As illustrated in Fig. 6, the (x_0, y_0) plane exhibits regions of distinct asymptotic behavior corresponding to different attractors of the system. The uniformly shaded blue region denotes trajectories converging toward a stable peri-

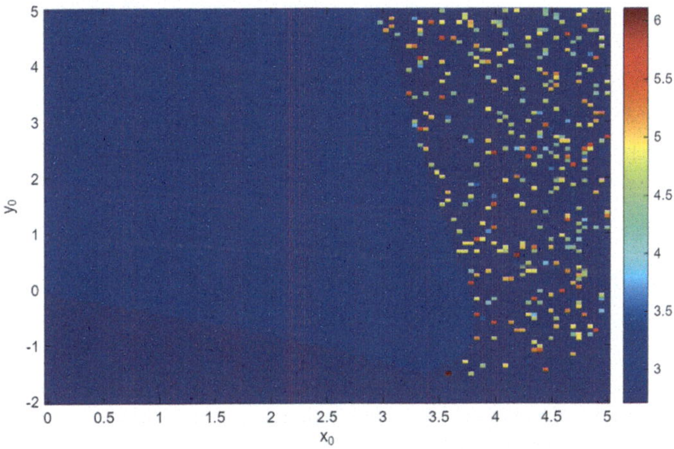

Fig. 6 Basin of attraction of the integer-order Strauss-based Skill Enhancement Chaotic Model in the (x_0, y_0) plane. The color intensity denotes the asymptotic regime corresponding to different initial conditions, highlighting multistability and sensitivity to initial states

odic or steady-state orbit, representing consistent and predictable enhancement in the learner's cognitive and practical performance levels. Conversely, the scattered multicolored zones observed at larger values of x_0 indicate multistability and irregular transitions between coexisting attractors. These discontinuous and fragmented domains suggest the existence of fractal basin boundaries, a hallmark of chaotic dynamics in nonlinear systems. Such structures confirm that small perturbations in the initial state can lead to significantly different dynamical evolutions, demonstrating strong sensitivity to initial conditions. This sensitivity captures the intricate nonlinear coupling among learning components–knowledge acquisition, skill application, and adaptive feedback–that govern the evolution of the integer-order model. The resulting topological complexity of the basin underscores the system's potential to exhibit both stable and chaotic learning trajectories depending on the learner's starting condition and the control parameter configuration.

4 Neural Network-Based Control for Integer-Order Strauss Skill System

In the Strauss-based Skill Enhancement Chaotic system (1), nonlinear interactions among theoretical knowledge $x(t)$, practical skills $y(t)$, and overall progress $z(t)$ generate intricate, often chaotic dynamics that are difficult to control using conventional strategies. To regulate these behaviors and steer the system along desired learning trajectories, a neural network-based adaptive control scheme is employed. Neural networks, inspired by the computational architecture of biological neurons,

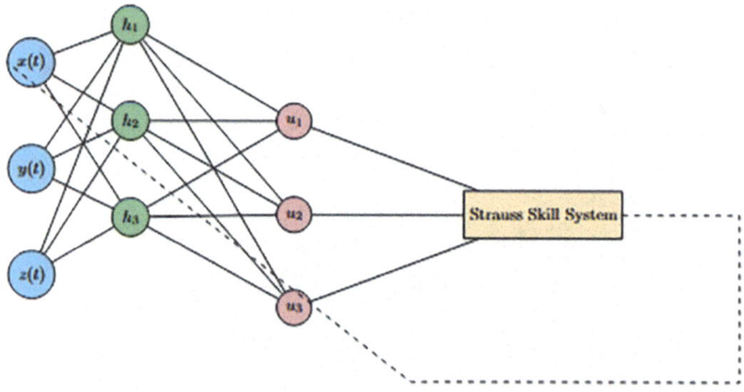

Fig. 7 Colorful compact neural network-based adaptive control for the integer-order Strauss skill system. Inputs (blue) feed into a hidden layer (green), producing control outputs (pink) that regulate the system (orange). Dashed arrow shows feedback of system states to the controller

are capable of approximating arbitrary nonlinear mappings and dynamically compensating for modeling uncertainties. Each neuron performs a nonlinear transformation of a weighted sum of its inputs, expressed as

$$a = \phi\left(\sum_{i=1}^{n} w_i x_i + b\right), \tag{2}$$

where x_i are inputs, w_i are adjustable synaptic weights, b is the bias, $\phi(\cdot)$ is a nonlinear activation function, and a is the neuron output.

In this framework, three adaptive control inputs $u_1(t)$, $u_2(t)$, and $u_3(t)$ are incorporated into the integer-order Strauss model to mitigate nonlinear uncertainties and synchronize the system states. The controlled dynamics are given by

$$\begin{aligned} \dot{x}(t) &= \alpha(y - x) + z\cos(x) + \mu x^2 + u_1(t), \\ \dot{y}(t) &= \beta x - yz + \gamma \sin(y) + \nu yz + u_2(t), \\ \dot{z}(t) &= \delta xy - \epsilon z + \eta \cos(yz) + \lambda xz + u_3(t), \end{aligned} \tag{3}$$

where the dot denotes the standard first-order derivative with respect to time.

The neural network controller in Fig. 7, adaptively estimates unknown nonlinearities and external disturbances while minimizing synchronization and tracking errors between the actual and desired learning trajectories. By continuously updating its internal parameters based on error feedback, the network ensures stable convergence of the system states within a bounded region corresponding to meaningful skill evolution. This methodology guarantees robustness against noise, parametric uncertainties, and time-varying interactions, providing a biologically inspired mechanism for stabilizing the integer-order Strauss skill dynamics.

4.1 Stability Analysis of the Skill Enhancement Model (1)

To ensure that the neural network controllers stabilize the integer-order Strauss-based Skill Enhancement Model, we present the following theorem.

4.1.1 Theorem on Stability

Theorem 2 (Stability of Neural Network-Controlled Skill Enhancement Model, Integer-Order) *Consider the controlled integer-order Strauss-based Skill Enhancement Model:*

$$\begin{aligned} \dot{x_d}(t) &= \alpha(y - x) + z\cos(x) + \mu x^2 + u_1(t), \\ \dot{y_d}(t) &= \beta x - yz + \gamma\sin(y) + \nu yz + u_2(t), \\ \dot{z_d}(t) &= \delta xy - \epsilon z + \eta\cos(yz) + \lambda xz + u_3(t), \end{aligned} \tag{4}$$

where $(\dot{\cdot})$ *denotes the standard derivative with respect to time, and* $x_d(t), y_d(t), z_d(t)$ *denote desired skill trajectories. Define the tracking errors:*

$$e_1 = x - x_d, \quad e_2 = y - y_d, \quad e_3 = z - z_d.$$

If the neural network-based control laws are designed as

$$u_i(t) = -\hat{f}_i(x, y, z) - k_i e_i, \quad k_i > 0, \quad i = 1, 2, 3, \tag{5}$$

where $\hat{f}_i(x, y, z)$ *approximates the unknown nonlinearities, then the closed-loop error system*

$$\dot{e}_i = f_i(x, y, z) - \hat{f}_i(x, y, z) - k_i e_i$$

is asymptotically stable in the classical Lyapunov sense. Consequently, the tracking errors $e_i(t)$ *converge to zero as* $t \to \infty$.

Proof Let the ideal unknown nonlinear skill dynamics be

$$\begin{aligned} f_1(x, y, z) &= \alpha(y - x) + z\cos(x) + \mu x^2, \\ f_2(x, y, z) &= \beta x - yz + \gamma\sin(y) + \nu yz, \\ f_3(x, y, z) &= \delta xy - \epsilon z + \eta\cos(yz) + \lambda xz. \end{aligned}$$

Define the neural network approximation errors:

$$\tilde{f}_i = f_i(x, y, z) - \hat{f}_i(x, y, z), \quad i = 1, 2, 3.$$

Consider the Lyapunov candidate function:

$$V = \frac{1}{2}\sum_{i=1}^{3} e_i^2 + \frac{1}{2}\sum_{i=1}^{3} \tilde{W}_i^T \Gamma_i^{-1} \tilde{W}_i, \tag{6}$$

where $\tilde{W}_i = W_i - W_i^*$ denotes the weight estimation error, and Γ_i is a positive definite adaptation gain matrix.

Taking the time derivative gives:

$$\dot{V} = \sum_{i=1}^{3} e_i \dot{e}_i + \sum_{i=1}^{3} \tilde{W}_i^T \Gamma_i^{-1} \dot{\tilde{W}}_i = \sum_{i=1}^{3} e_i(\tilde{f}_i - k_i e_i) + \sum_{i=1}^{3} \tilde{W}_i^T \Gamma_i^{-1} \dot{\tilde{W}}_i.$$

Choose the neural network weight update law as

$$\dot{\hat{W}}_i = \Gamma_i e_i \varphi_i(x, y, z),$$

where $\varphi_i(\cdot)$ is the activation vector. Then

$$\tilde{W}_i^T \Gamma_i^{-1} \dot{\tilde{W}}_i = -e_i \tilde{f}_i,$$

so the Lyapunov derivative simplifies to

$$\dot{V} = -\sum_{i=1}^{3} k_i e_i^2 \leq 0.$$

Since $\dot{V} \leq 0$ and $k_i > 0$, $V(t)$ is non-increasing, implying boundedness of the errors. By standard Lyapunov theory, the equilibrium point $e = 0$ is globally asymptotically stable. Hence, the neural network-based controller ensures that the tracking errors converge to zero, guaranteeing stable skill enhancement dynamics. □

4.2 Simulation Results

The Strauss-based Skill Enhancement Chaotic Model (4), integrated with the RBF neural network controller (5) and the corresponding adaptive laws, was numerically simulated in MATLAB. The uncontrolled system exhibits intrinsic chaotic behavior due to strong nonlinear couplings among theoretical knowledge $x(t)$, practical skills $y(t)$, and overall progress $z(t)$. The application of adaptive neural network control effectively regulates these dynamics and the convergence of the tracking error as depicted in Fig. 8. Figure 9, further illustrates the temporal evolution of $x(t)$, $y(t)$, and $z(t)$ under control, highlighting rapid convergence toward desired reference trajectories.

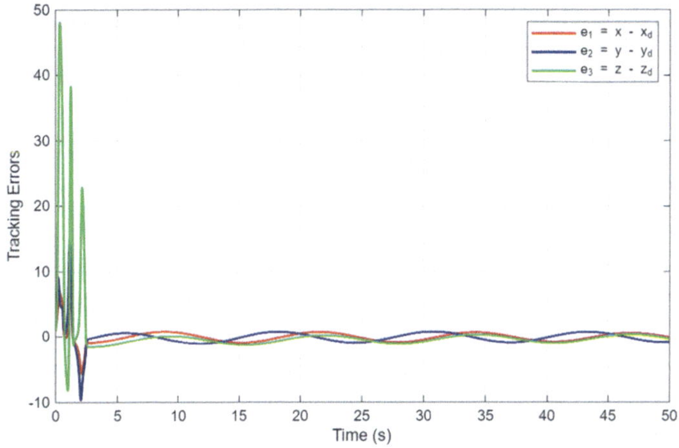

Fig. 8 Error States under the controller of the integer-order Strauss-based Skill Enhancement Chaotic Model

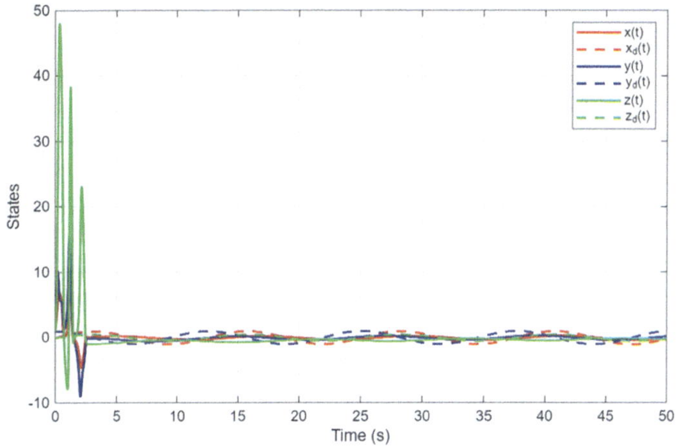

Fig. 9 Controlled States under the controller of the integer-order Strauss-based Skill Enhancement Chaotic Model

The controller continuously estimates unknown nonlinearities and compensates for perturbations by minimizing the tracking errors $e_i(t) = x_i(t) - x_{i,d}(t)$ in real time. This adaptive feedback ensures that the system trajectories are steered toward stability despite the chaotic intrinsic dynamics. Specifically, stabilization of $x(t)$ reflects controlled accumulation of theoretical knowledge, avoiding excessive fluctuations, while convergence of $y(t)$ demonstrates effective acquisition of practical skills with reduced cognitive overload. The overall progress $z(t)$ reaching steady values indicates coherent integration of theoretical and practical components, consistent with structured skill enhancement objectives.

These results confirm that the proposed neural network-based adaptive control strategy not only suppresses chaotic oscillations inherent in the system but also guarantees systematic error reduction and stable learning dynamics. Consequently, the framework provides a robust, biologically inspired mechanism for guiding skill acquisition along well-defined trajectories, ensuring both theoretical and practical aspects of learning are simultaneously optimized.

5 Conclusion

This chapter provides a comprehensive interdisciplinary framework that integrates the theory of chaos , mathematical modelling, and neural network-based control to advance the implementation of skill-oriented higher education. Further, the complex nonlinear interactions among theoretical knowledge, practical skill acquisition and overall learning progress were effectively demonstrated by developing a chaotic dynamical model. In addition to this, the dissipativity analysis confirmed the bounded nature of system trajectories, while the phase-space visualizations, Lyapunov exponents, and bifurcation diagrams revealed deterministic chaos in the learning dynamics. Furthermore, the basin of attraction analysis demonstrated the system's sensitivity to initial learning conditions, underscoring the necessity of adaptive strategies for ensuring consistent educational outcomes.

To stabilize these chaotic behaviors, an adaptive neural network control scheme was employed, enabling the convergence of system states toward desired trajectories despite nonlinear perturbations and parameter uncertainties. The proposed control approach ensured robust stabilization and systematic enhancement of learning performance. Overall, the results highlight a novel pathway for designing adaptive, resilient, and data-driven higher education systems. This work provides valuable insights for educators and policymakers seeking to optimize learning trajectories, minimize instability, and align academic curricula with the evolving demands of employability and lifelong learning.

Competing Interests The authors confirm that there are no relevant financial or personal conflicts of interest associated with this chapter.

Ethics Approval Ethics approval was not applicable for this chapter.

Acknowledgements I express my sincere gratitude to Professor Ayub Khan for his guidance, support, and invaluable insights, which were essential in completing this research.

References

1. Garnjost, P., Brown, S.M.: Undergraduate business students' perceptions of learning outcomes in problem based and faculty centered courses. Int. J. Manag. Educ. **16**(1), 121–130 (2018)

2. Andrews, J., Higson, H.: Graduate employability, 'soft skills' versus 'hard' business knowledge: a European study. Eur. J. High. Educ. **33**(4), 411–422 (2008)
3. Carnevale, A.P., Smith, N., Strohl, J.: Recovery: Job Growth and Education Requirements Through 2020. Georgetown University Centre on Education and the Workforce (2013)
4. Clarke, M.: Rethinking graduate employability: the role of capital, individual attributes, and context. STUDS **43**(11), 1923–1937 (2018)
5. Marope, P.T.M.: Quality and development-relevant education and learning: setting the stage for the education 2030 Agenda. Prospects **46**(1), 1–3 (2016)
6. Valiente, O.: The OECD skills strategy and the education agenda for development. Int. J. Educ. Dev. **39**, 40–48 (2014)
7. Tymon, A.: The student perspective on employability. Stud. High. Educ. **38**(6), 841–856 (2013)
8. Jackson, D.: Employability skill development in work-integrated learning: barriers and best practice. Stud. High. Educ. **40**(2), 350–367 (2015)
9. Finch, D.J., Hamilton, L.K., Baldwin, R., Zehner, M.: An exploratory study of factors affecting undergraduate employability. Educ. Train. **55**(7), 681–704 (2013)
10. Leest, B., Wolbers, M.H.: Critical thinking, creativity and study results as predictors of selection for and successful completion of excellence programmes in Dutch higher education institutions. Eur. J. High. Educ. **11**(1), 29–43 (2021)
11. Jackson, D., Bridgstock, R.: Evidencing student success and career outcomes among business and creative industries graduates. High. Educ. Policy Manag. **41**(5), 451–467 (2019)
12. Romanova, O.: Embedding employability skills into vocational education and training: What works best for students' self-evaluation and aspirations? Teach. Learn. Grad. Employab. **13**(1), 20–36 (2022)
13. Qenani, E., MacDougall, N., Sexton, C.: An empirical study of self-perceived employability. J. Educ. Bus. **89**(4), 215–224 (2014)
14. Andrews, J., Higson, H.: Graduate employability, 'soft skills' versus 'hard' business knowledge: a European study. Eur. J. High. Educ. **33**(4), 411–422 (2008)
15. Koopmans, M.: Education is a complex dynamical system: challenges for research. J. Exp. Educ. **88**(3), 358–374 (2020)
16. Vulic, J., Jacobson, M.J., Levin, J.A.: Exploring education as a complex system: computational educational research with Multi-Level Agent-Based modeling. Educ. Sci. **14**(5), 551 (2024)
17. Lippolis, D.: Thermodynamics of chaotic relaxation processes. Phys. Rev. E **110**(2) (2024)
18. Wang, Z., Wei, Z., Sun, K., He, S., Wang, H., Xu, Q., Chen, M.: Chaotic flows with special equilibria. Eur. Phys. J. Spec. Top. **229**(6), 905–919 (2020)
19. Cang, S., Wang, Z., Chen, Z., Jia, H.: Analytical and numerical investigation of a new Lorenz-like chaotic attractor with compound structures. Nonlinear Dyn. **75**(4), 745–760 (2014)
20. Chen, G.D.X.: From Chaos to Order: Methodologies, Perspectives, and Applications. World Scientific (2015)
21. Vaidya, T., Chotibut, T., Piliouras, G.: Broken detailed balance and non-equilibrium dynamics in noisy social learning models. Phys. A Stat. Mech. Appl. **570**, 125818 (2021)
22. Ding, D., Qian, X., Hu, W., Wang, N., Liang, D.: Chaos and Hopf bifurcation control in a fractional-order memristor-based chaotic system with time delay. Eur. Phys. J. Plus. **132**(11), 447 (2017)
23. Mehlig, B.: Machine Learning with Neural Networks: An Introduction for Scientists and Engineers, Cambridge University Press (2021)
24. Zhang, Y., Tao, G., Chen, M.: Adaptive neural network based control of noncanonical nonlinear systems. IEEE Trans. Neural Netw. Learn. Syst. **27**(9), 1864–1877 (2015)
25. Kumar, P., Lee, T.H., Ertruk, V.S.: A fractional-order multi-delayed bicyclic crossed neural network: stability, bifurcation, and numerical solution. Neural Netw. **188**, 107436 (2025)
26. Kumar, P., Lee, T.H., Ertruk, V.S.: A novel fractional-order neutral-type two-delayed neural network: stability, bifurcation, and numerical solution. Math. Comput. Simul. **232**, 245–260 (2025)
27. Kumar, P., Lee, T.H., Ertruk, V.S.: A novel two-delayed tri-neuron neural network with an incomplete connection. Nonlinear Dyn. **112**, 20269–20293 (2024)

28. Johansyah, M.D., Sambas, A., Hannachi, F., Hamidzadeh, S.M., Rusyn, V., Hidayanti, M., Foster B, Rusyaman, E.: Dynamics and stabilization of chaotic monetary system using radial basis function neural network control. Mathematics **12**(24), 3977 (2024)
29. Cheng, L., Wang, Z., Jiang, F., Li, J.: Adaptive neural network control of nonlinear systems with unknown dynamics. Adv. Space Res. **67**(3), 1114–1123 (2021)
30. Chen, X., Cao, J., Qiu, J., Alsaedi, A., Alsaadi, F.E.: Adaptive control of multiple chaotic systems with unknown parameters in two different synchronization modes. Adv. Differ. Equat. **2016**(1), 231 (2016)
31. Liu, J., Liu, K., Liu, S.: Adaptive control for a class of nonlinear complex dynamical systems with uncertain complex parameters and perturbations. PLoS ONE **12**(5), e0175730 (2017)
32. Huang, C., Liu, Z., Chen, C.P., Zhang, Y.: Adaptive fixed-time neural control for uncertain nonlinear multiagent systems. IEEE Trans. Neural Netw. Learn. Syst. **34**(12), 10346–10358 (2022)
33. Akmansoy, V., Kartal, S.: Chaos theory and its application to education: Mehmet Akif Ersoy University Case. Educ. Sci. Theory Pract. **14**(2), 510–518 (2014)
34. Hwang, G.J., Sung, H.Y., Chang, S.C., Huang, X.C.: A fuzzy expert system-based adaptive learning approach to improving students' learning performances by considering affective and cognitive factors. Comp. Educ. Artif. Intel **1**, 100003 (2020)
35. Tavakoli, M., Faraji, A., Molavi, M., Mol, T., Kismihók, S., G.: Hybrid human-AI curriculum development for personalised informal learning environments. In: 12th ILAKC, pp. 563–569
36. Horwitz, L., Strauss, Y.: Unstable System. Springer (2020)

Predicting the Ebola Virus Outbreak in Guinea by Combining Neural Network Models with a Deterministic Model

Delly Kassala, Ebenezer Maka Maka, Yannick Malong, Benjamin Garga, Daouda Hassana Daouda, and Hamadjam Abboubakar

Abstract Between 2014 and 2016, the *filoviridae* virus, which causes Ebola virus disease, caused severe epidemics, especially in West Africa. Many mathematical models were created in response to this health disaster in order to study and forecast the epidemic's spread. The goal of this research is to use artificial intelligence techniques, specifically deep learning, to forecast new Ebola virus cases. Long Short-Term Memory (LSTM) and Gated Recurrent Unit (GRU) techniques will be used to estimate the dynamic parameters of the fundamental model, which is a Susceptible-Exposed-Infectious-Recovered-Death-Bacteria type model. The parameters obtained are then used to simulate the model and forecast new infections and deaths in Guinea. To evaluate the capability of the Long Short-Term Memory and Gated Recurrent Unit approaches in predicting epidemics, the results were compared using standard error measures such as the root mean square error, mean absolute percentage error, and R^2. The results demonstrate that the Long Short-Term Memory and Gated Recurrent Unit approaches have encouraging promise to increase the precision of epidemiological forecasts, especially for extremely deadly illnesses like Ebola.

D. Kassala · B. Garga · D. Hassana Daouda
University of Ngaoundéré, University Institute of Technology, Laboratory of Analysis, Simulation and Tests, Ngaoundéré, Cameroon

E. Maka Maka · Y. Malong
Department of Computer Engineering and Telecommunications, University of Douala, National Higher Polytechnic School, Douala, Cameroon

H. Abboubakar (✉)
Department of Computer Engineering, University of Ngaoundéré, University Institute of Technology, Ngaoundéré, Cameroon
e-mail: h.abboubakar@gmail.com

Laboratory of Analysis, Simulation and Tests, University of Ngaoundéré, University Institute of Technology, Ngaoundéré, Cameroon

Department of Applied Mathematics and Computer Science, University of Ngaoundéré, School of Geology and Mining Engineering, Meiganga, Cameroon

L. Kaur and P. Kumar (eds.), *Neural Networks: Theory, Algorithms, Simulation, and Applications*, Studies in Computational Intelligence 1273,
https://doi.org/10.1007/978-3-032-18750-5_9

1 Introduction

Between 2014 and 2016, West Africa had one of the most severe Ebola outbreaks since the virus was first detected. Guinea, Liberia, and Sierra Leone were the most affected nations, with 28,618 confirmed cases and 11,310 fatalities [1]. This disease has caused a major public health crisis, with thousands of deaths. The Ebola virus, a member of the Filoviridae family, is well-known for its capacity to produce pathogenic agents and release this pathogens in the body, with a mortality rate of up to 90% in some cases [2].

Ebola Virus Disease (EVD), named for the Ebola River that flows near Yambuku in the Democratic Republic of the Congo, was first discovered in 1976 in Africa, specifically in what is now South Sudan and the DRC. It has caused approximately 318 cases and 280 fatalities, as well as 284 cases and 151 deaths [3, 4]. The bat is the main viral reservoir in tropical regions. It spreads among people after being brought to humans by large apes. The virus can only be transmitted by direct contact with bodily fluids such blood, vomit, feces, sweat, or saliva by ill or symptomatic people. It should be noted that contact can occur directly between people or indirectly through non-disinfected liquids or fluid-contaminated objects. Fever, tiredness, headache, sore throat, and muscle discomfort are the most obvious symptoms of EVD. They are followed by internal and external bleeding, rash, diarrhea, and nausea. Between two and twenty-one days can pass between a viral infection and the onset of symptoms [5]. Given the severity of the pandemic, efforts must be made to limit its spread, with a focus on immunization campaigns, medical treatments, and other strategies. The erratic behavior of animals that serve as viral reservoirs has made it challenging to manage the pandemic in spite of these efforts.

Since Sir Ronald Ross's groundbreaking research on malaria in the 1890s and 1900s, the study of direct and/or indirect communicable diseases has become increasingly important. Numerous models have now been created to study the spread of illnesses like cancer, pneumonia, TB, HIV/AIDS, Hepatitis B, COVID-19, and Ebola [6–17]. In order to estimate the spread of infectious diseases while accounting for a number of factors, including population characteristics and migrations, mathematical models—especially epidemiological or compartmental models—are essential. They provide methods for managing epidemics and assessing medical needs. In order to provide political decision makers with accurate information to help them develop and plan effective policies that lessen the health and socioeconomic implications, epidemiological models can be used to predict the course of an epidemic and simulate alternative scenarios [18].

Several mathematical models have been developed to control the spread of the Ebola virus, considering direct contact with infected individuals or objects [9, 19–28]. Two mathematical models of Ebola virus outbreaks were studied in [29]. Data from Guinea, Liberia, and Sierra Leone were used in the simulation for the Susceptible-Exposed-Infectious-Recovered ($SEIR$) model. The generalized epizootic model that explains the dynamics of the transmission of Ebola virus disease in bat populations is the subject of the work of [22]. Saturated incidence, classical bilinear incidence,

Beddington-DeAngelis functional response, Hattaf-Yousfi functional response, and Crowley-Martin functional response are among the functions their model takes into consideration when simulating bat contamination. As a result, [30] presents a model of the Ebola virus outbreak that incorporates both vaccination and quarantine. This work is unique in that it simulates the aleatoric process of the Ebola virus's spread using the Markov Chain Monte Carlo (MCMC) technique. Euler's approach is then applied to the Monte Carlo simulation, which has improved reality observation and allowed for a more accurate understanding of how the Ebola virus spreads in crowds.

These mathematical models of the Ebola disease take into account direct contact with infected persons or things, emphasizing the relevance of repeated exposure in Sub-Saharan African nations. In a recent study [19], a traditional compartmental model of the SEIR epidemic was expanded to include convex incidence rates, death rates, and pathogen classifications. This is the first attempt to incorporate this element into the dynamics of the Ebola model. The disease-free equilibrium was shown to be locally and globally stable, as was the presence of a unique endemic equilibrium and its global stability. A model calibration was performed to estimate some values of the model parameters, and numerical simulations were authorized.

One of the problems in epidemiological modeling is estimating the values of the model parameters, which can change depending on a variety of factors such as human behavior, medical interventions, quarantines, and even environmental circumstances. Estimating these characteristics is critical if we want to make accurate predictions about the evolution of the pandemic in real time [31]. For parameter estimation, a variety of traditional methods based on classical statistical approaches are utilized, including least squares, Bayesian inference, Kalman filtering, and others. These approaches may lack accuracy due to a variety of factors, including unpredictability of the data, computational complexity, and sensitivity to errors. This imprecision can lead to incorrect forecasts and unproductive decision-making, worsening the epidemic's effects. Furthermore, some of these approaches assume that the parameters of infectious disease transmission models are constant over time, which is not true for many models. To address the problems and limits of classic parameter estimation approaches, deep learning techniques are effective because of their ability to capture non-linear correlations and complicated dynamics in epidemiological data [32, 33]. Machine Learning algorithms may be made more adaptable to changes in data during epidemic situations [34–36].

The goal of our research is to use deep learning techniques, namely the Long Short Time Memory (LSTM) and Gated Recurrent Unit (GRU) architectures, to forecast new instances of infected and deceased persons using the SEIRD-B model for Ebola in Guinea studied in [19]. In fact, in this previous work, we used the parameter estimation method called nonlinear least squares. The major drawback of this method is that it assumes that the parameter values remain fixed over time. To overcome this problem, we will use machine learning algorithms. Specifically:

(i) Presentation of the SEIRD-B study model with differential equations describing the transition between compartments;
(ii) Euler discretisation of these equations in order to derive the expressions of the parameters that will be calculated and constitute the data set that will be normalised and trained with LSTM and GRU;
(iii) Simulation of the model with the parameters obtained from LSTM and GRU and evaluation of performance with RMSE, MAPE and R2 illustrating the main conclusions via graphs comparing predictions with actual data;
(iv) And finally, direct training on epidemiological data with LSTM and GRU compared to actual data in order to evaluate accuracy and performance with the prediction of I and D using the parameters predicted with LSTM and GRU.

The organization of the work is as follows: Sect. 2 is devoted to the methodology used to address the problem. Section 3 presents the obtained results, while Sect. 4 is devoted to discussion. Finally, Sect. 5 concludes the study.

2 Methods

2.1 SEIRD-B Model Formulation

The model therefore offers an interesting potential for improving the robustness and precision of the estimations of epidemiological parameters. The SEIRD-B model extends the traditional SEIR model by including fatalities and infections in the environment. The model is split into six compartments (sub-populations):

(1) **Susceptible individuals (S)**: This compartment depicts persons who are prone to become infected with the illness. They have not yet been exposed to the pathogen and may get the illness if they come into touch with sick people;
(2) **Exposed individuals (E)**: Individuals in this compartment have been exposed to the pathogen but are not now infectious. They are in a latent phase, which means they have the virus but cannot transmit it to others. This compartment models the time between exposure and infectivity;
(3) **Infectious individuals (I)**: This compartment contains individuals who are currently infected and capable of spreading the disease to others. They have completed the incubation phase and are now a viable source of disease proliferation;
(4) **Recovered individuals (R)**: Individuals in this compartment were infected, but they recovered. They have acquired temporary or permanent immunity to the illness and can no longer spread it;
(5) **Deaths (D)**: This division reflects those who died as a result of the disease. Unlike in other infectious models, death individuals continue to play a role in the infection's dynamics;
(6) **Free viruses in the environment (B)**: This compartment represents the concentration of the virus in the environment.

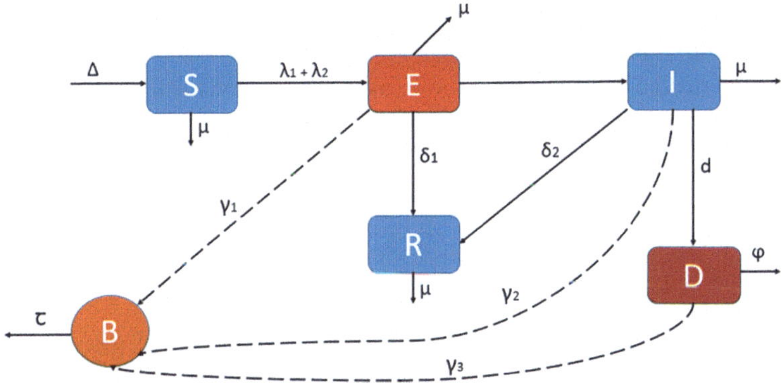

Fig. 1 Representation of the SEIRD-B compartmental model [19]

For our case study, we use the SEIRD-B model studied by Abboubakar et al. [19]. This model subdivides the study population into six compartments: Susceptible (S), Exposed (E), Infected (I), recovered (R), Deceased (D) and free pathogens in the environment (B). From Fig. 1, we deduce the following system of ordinary differential equations:

$$\begin{cases} \frac{dS}{dt} &= \Delta - \lambda S - \beta_2 SD(1 + \alpha_4 D) - \mu S, \\ \frac{dE}{dt} &= \lambda S + \beta_2 SD(1 + \alpha_4 D) - (b_1 + \delta_1 + \mu)E, \\ \frac{dI}{dt} &= b_1 E - (\delta_2 + d + \mu)I, \\ \frac{dR}{dt} &= \delta_1 E + \delta_2 I - \mu R, \\ \frac{dD}{dt} &= dI - \phi D, \\ \frac{dB}{dt} &= \gamma_1 E + \gamma_2 I + \gamma_3 D - \tau B. \end{cases} \tag{1}$$

where $\lambda = \beta_1 B(1 + \alpha_1 B) + \beta_1 \psi_1 E(1 + \alpha_2 E) + \beta_1 \psi_2 I(1 + \alpha_3 I)$ denotes the force of infection. The description of the model parameters is presented below:

- Δ: Recruitment rate (birth or immigration) in the susceptible population;
- β_1: Transmission rate of pathogens from the environment (B) to the susceptible (S);
- $\alpha_1, \alpha_2, \alpha_3, \alpha_4$: rates of disease transmission after double exposure,
- ψ_1, ψ_2: Factors modifying the rate of transmission from the exposed (E) and infected (I) to the susceptible (S);
- β_2: Rate of transmission of pathogens from the dead (D) to the susceptible (S);
- μ: Natural mortality rate for each compartment;
- b_1: Rate of progression from exposed (E) to infected (I);
- δ_1: Recovery rate for exposed individuals (E);
- δ_2: Recovery rate for infectious individuals (I);
- d: Mortality rate due to infection;
- ϕ: Burial rate of the deceased;

- $\gamma_1, \gamma_2, \gamma_3$: Contribution rate of exposed (E), infectious (I) and dead (D) individuals to virus proliferation in the environment;
- τ: Natural decomposition rate of the virus in the environment.

For positivity and boundedness of solutions, we have the following result:

Lemma 1 ([19]) *Let* $X(t) = (S(t), E(t), I(t), R(t), D(t), B(t))$ *and* $N = S(t) + E(t) + I(t) + R(t) + D(t)$.

1. *With nonnegative initial condition* $X(0)$, *each solution* X *of* (1) *is nonnegative* $\forall t > 0$.
2. *The following set* $\Upsilon = \left\{X(t) \in \mathbb{R}_+^6 : N \leq \frac{\Delta}{\mu};\ D(t) \leq \frac{d\Delta}{\phi\mu};\ B(t) \leq \frac{\Delta}{\mu\tau}\left(\gamma_1 + \gamma_2 + \frac{d\gamma_3}{\phi}\right)\right\}$ *is attracting for model* (1).

2.2 Basic Reproduction Number $\mathcal{R}_0$

The basic reproduction number $\mathcal{R}_0$ is a key indicator for measuring the potential spread of an infectious disease in a population, representing the average number of new infections produced by a case in a population that is totally susceptible and unaffected by external forces. For the SEIRD-B model, R_0 is calculated as described in [19]: The system (1) admits $\mathcal{E}_0 = (S_0, 0, 0, 0, 0, 0) = \left(\frac{\Delta}{\mu}, 0, 0, 0, 0, 0\right)$ as the disease-free equilibrium point. Applying the next-generation method (NGM) described in [37], we define the matrix of infection terms $\mathcal{F}$ and the matrix of transfer terms $\mathcal{V}$ as follows:

$$\mathcal{F} = \begin{pmatrix} S_0\beta_1\psi_1 & S_0\beta_1\psi_2 & S_0\beta_2 & S_0\beta_1 \\ 0 & 0 & 0 & 0 \\ 0 & 0 & 0 & 0 \\ 0 & 0 & 0 & 0 \end{pmatrix} \quad \text{and} \quad \mathcal{V} = \begin{pmatrix} -k_2 & 0 & 0 & 0 \\ b_1 & -k_3 & 0 & 0 \\ 0 & d & -\varphi & 0 \\ \gamma_1 & \gamma_2 & \gamma_3 & -\tau \end{pmatrix}, \tag{2}$$

where $k_2 = b_1 + \delta_1 + \mu$ and $k_3 = d + \delta_2 + \mu$.

The basic reproduction is defined as the spectral radius of the next-generation matrix $-\mathcal{F}\mathcal{V}^{-1}$, that is

$$\mathcal{R}_0 = \rho(-\mathcal{F}\mathcal{V}^{-1}) = \mathcal{R}_E + \mathcal{R}_I + \mathcal{R}_D + \mathcal{R}_B. \tag{3}$$

with
$\mathcal{R}_E = \frac{\psi_1\beta_1}{k_2}\frac{\Delta}{\mu}$, $\mathcal{R}_I = \frac{\psi_2 b_1\beta_1}{k_2k_3}\frac{\Delta}{\mu}$, $\mathcal{R}_D = \frac{b_1\beta_2 d}{k_2k_3\varphi}\frac{\Delta}{\mu}$, and $\mathcal{R}_B = \frac{\beta_1\left[b_1\gamma_3 d + (\gamma_1 k_3 + b_1\gamma_2)\Delta\varphi\right]}{\mu k_2k_3\tau\varphi}$.

There are three scenarios that indicate possible regression or spread of the disease depending on the value of $\mathcal{R}_0$:

(a) $\mathcal{R}_0 = 1$: This indicates the equilibrium reached by the disease. Each infected individual can infect exactly one new person;

(b) $\mathcal{R}_0 < 1$: each infected individual infects less than one new person, which explains why the disease fades over time;
(c) $\mathcal{R}_0 > 1$: each infected individual infects more than one new person, which means that the disease will spread among individuals, leading to an outbreak or epidemic.

2.3 *Challenges of Parameter Estimation*

Since the model parameter values have a direct effect on the concordance of the results, it is crucial for the solution of our SEIRD-B model, to estimate these parameters. As mentioned in the introduction, there are a number of methods used to estimate model parameters, such as the use of evolutionary algorithms, Bayesian inference, maximum similarity, and least squares. While some parameters are time-dependent and may vary for a variety of reasons, such as government policies like quarantine, curfews, vaccinations, and medical effects, the parameters estimated with these techniques according to the models may respond to static trends in the epidemic. In the case of sequential data, however, the methods cited above cannot be applied for the estimation of parameters for a long duration because they are only for the short term. The proposed solution employs deep learning algorithms to estimate specific dynamic parameters, including the transmission rate, the death rate resulting from an epidemic, and the degradation rate of environmental pathogens.

2.4 *Deep Learning*

With the increase in epidemics and the requirement for precise forecasts to handle health emergencies, deep learning has emerged as a crucial tool in epidemiological modeling. To improve forecasting, more complex dynamics can be captured through the incorporation of deep learning approaches into epidemiological modeling. We suggest a deep learning approach based on the LSTM and GRU architectures for estimating the parameters of the SEIRD-B model in order to get beyond the drawbacks of traditional parameter estimation strategies.

2.5 *Artificial Neural Networks (ANN)*

Artificial Neural Networks (ANNs) are deep learning systems designed to mimic the structure and function of biological neurons. They consist of interconnected layers of artificial neurons, where each neuron processes input signals and passes the result to the next layer. The different layers are:

- Input layers that receive the raw data;
- Hidden layers that apply non-linear transformation methods to process information;
- Output layers giving the final results.

2.6 Recurrent Neural Networks (RNN)

Recurrent neural networks are part of artificial neural networks, which have memories that allow them to learn while retaining past information and using it to make future predictions. The RNN module is made up of values defined as follows:

- Hidden state that saves past information that is updated at each stage of the sequence;
- Sequential inputs, which characterise the entry of data in sequence into the network;
- Sequential output where each hidden state produces an output at each step, influenced by the current input and the previous hidden state.

Their architecture can be summed up in three main elements:

- x_t: represents entry at time t;
- h_t: represents the cell's hidden memory at time t;
- $W_x(t)$: is the weight matrix associated with x at time t;
- $W_y(t)$: is the weight matrix associated with h_{t-1} at time t.

The RNN mechanism is based on the combination of the current input x_t and the previous hidden state h_{t-1}, using the weight matrices $W_x(t)$ and $W_y(t)$ with bias b. The tanh activation function normalizes the output between -1 and +1, introducing non-linearity to handle complex data. The new hidden state h_t is then used as the previous state for the next time step $t + 1$, allowing a dynamic memory to be maintained. RNNs are used for the predicting infected cases in an epidemic. Priyanka et al. on Covid-19 used RNN and LSTM for the predicting infected cases. They found that the LSTM model gives a promising result of 98% [38]. RNNs have a good ability to capture complex relationships in sequential data and are applicable to data with short sequences. However, it should be noted that RNNs have difficulty capturing long-term sequential dependencies due to gradient loss, and have a short and limited memory capacity, hence the interest in using LSTM and GRU, which are variants of RNNs that provide a solution to these limitations [39, 40]. Figure 2 depicts RNN cell.

$$h_t = tanh(W_{h(t)} * h_{t-1} + W_{x(t)*x_t} + b)$$

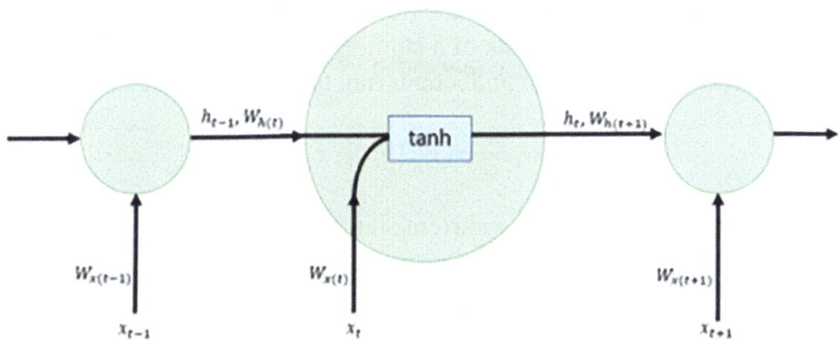

Fig. 2 A RNN cell [41]

2.7 *Long Short Term Memory (LSTM)*

To overcome the limitations of RNNs due to vanishing gradient problems limiting the learning of long-term data dependencies, LSTM models, proposed in 1997 by Sepp Hochreiter and Jürgen Schmidhuber, are capable of processing and predicting long-term sequential data. The two researchers introduced special cells that have the capacity to retain information over long periods. An LSTM cell is composed as follows:

(1) **Current state of the cell**: it is a long-term memory that allows the whole sequence to be run through in different time steps without major modification;
(2) **Gates**: LSTM uses three gates to control information:

  (i) **Forget Gate**: This gate determines what information the state of the cell should forget. This is done by a mathematical operation on the activation vector and the previous state C_{t-1}. From these two operations on C_{t-1}, we obtain the current state C_t. The forget gate uses a sigmoid function for the output, where 0 means forget the information and 1 means keep it;

$$F_t = \sigma(W_f[h_{t-1}, x_t] + b_f). \quad (4)$$

  (ii) **Input Gate**: This gate decides what new information will be kept in the current state. It consists of a sigmoid function to decide which values to update.and a tanh function to create vectors of new candidates to add. It takes into account the current input x_t and the previous state of the memory cell h_{t-1} and generates an activation vector for the cell C_{t-1}.

$$i_t = \sigma(W_i x_t + U_i h_{t-1} + b_i),$$
$$\tilde{C}_t = \tanh(W_C x_t + U_C h_{t-1} + b_C). \tag{5}$$

(iii) **Output Gate**: This gate takes the decision on the output of the current cell state information. It consists of a sigmoid function to determine which cell state to pass to the output and a tanh function to modulate the values:

$$O_t = \sigma(W_o[h_{t-1}, x_t] + b_o). \tag{6}$$

It uses the inputs x_t and the current state of the memory cell C_t to generate an activation vector which represents the output of LSTM:

$$h_t = O_t * \tanh(C_t). \tag{7}$$

The combination of these three gates allows LSTM to effectively manage long-term dependencies. During backpropagation, LSTM keeps the information flow constant over time, avoiding the problem of gradient loss and enabling stable and robust learning.

To predict the parameters with the LSTM, we define the architecture as follows:

- Input: The theoretically calculated parameters are pre-processed into sequential input data;
- LSTM Layer: We define the LSTM layers and cell adjusted according to the performance of our model on the validation set;
- Dense layer: this layer is used to produce the results of the predicted parameters at the output;
- Activation: to capture the non-linearity of the data, we use the ReLu activation function for the dense layer and Sigmoid for the output;
- Dropout: to avoid overfitting, it is necessary to use a dropout regularization.

Figure 3 depicts cell of LSTM.

2.8 GRU

A variant of LSTM that combines the functions of the LSTM input and output gates, reducing the structure and number of parameters. It consists of:

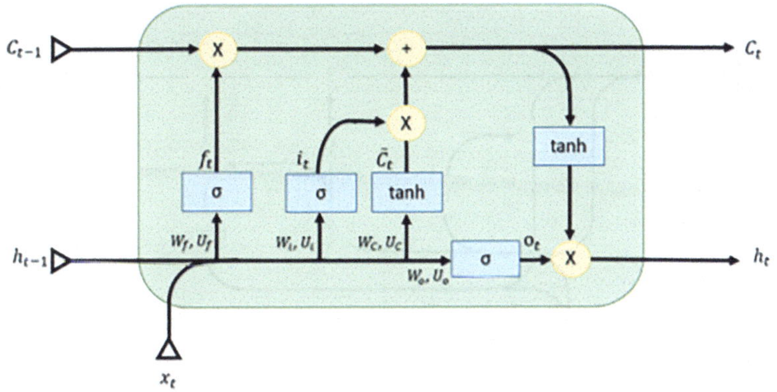

Fig. 3 LSTM cell [41]

Reset Gate:

It works in a similar way to the LSTM forget gate, controlling how much of the previous state should be deleted or forgotten.

$$R_t = \sigma(W_r \cdot [h_{t-1}, x_t] + b_t). \tag{8}$$

update Gate:

it combines the input and output gates of LSTM, determining how much of the previous state is maintained and how much new information is added.

$$Z_t = \sigma(W_z \cdot [h_{t-1}, x_t] + b_z). \tag{9}$$

Activation candidate:

generates the new candidate values to be added to the

$$\tilde{t} = \tanh(W \cdot [r_t * h_{t-1}, x_t] + b). \tag{10}$$

Updates are made by interpolating between the previous state and the new candidates:

$$h_t = (1 - Z_t) * h_{t-1} + Z_t * \tilde{t}. \tag{11}$$

To predict the parameters with the GRU, we define the same architecture as for LSTM:

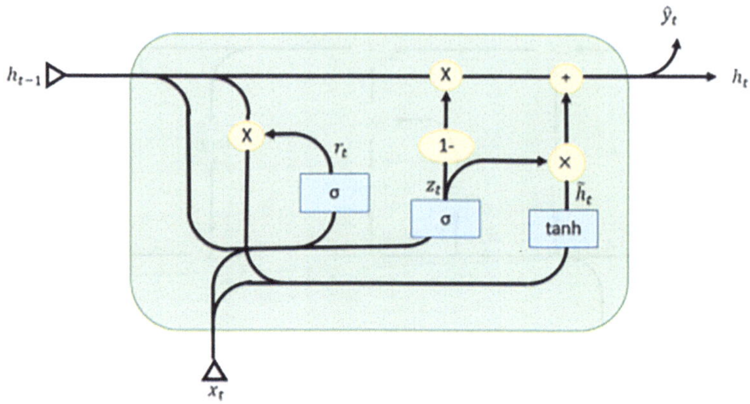

Fig. 4 GRU cell [41]

- Input: The theoretically calculated parameters are pre-processed into sequential input data;
- GRU Layer: We define the layers and cell of LSTM adjusted according to the performance of our model on the validation set;
- Dense layer: this layer is used to produce the results of the predicted parameters at the output;
- Activation: to capture the non-linearity of the data, we use the ReLu activation function for the dense layer and Sigmoid for the output;
- Dropout: to avoid overlearning it is necessary to use a Dropout regularization.

Figure 4 represents the GRU cell.

2.9 SEIRD-B-LSTM/GRU Algorithm

In this section we present the architecture of the methodology for estimating the dynamic parameters of the SEIRD-B model with LSTM and GRU and the discretizations of the model with these estimated parameters, illustrated by Fig. 5. All sub-sections are set out in the flow chart below.

The methodology adopted in this work is established as follows:

1. Presentation of the SEIRD-B study model with differential equations describing the transition between compartments;
2. Euler discretizations of these equations in order to derive the expressions of the parameters that will be calculated and constitute the data set that will be normalized and trained with LSTM and GRU;
3. Simulation of the model with the parameters obtained from LSTM and GRU and evaluation of performance with RMSE, MAPE and R^2 illustrating the main conclusions via graphs comparing predictions with actual data;

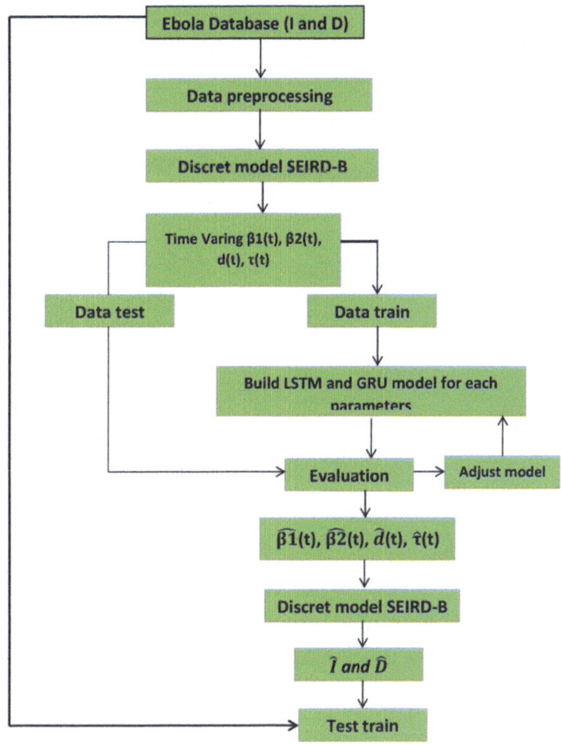

Fig. 5 SEIRD-B-LSTM/GRU model flow chart

4. And finally, direct training on epidemiological data with LSTM and GRU compared to actual data in order to evaluate accuracy and performance with the prediction of I and D using the parameters predicted with LSTM and GRU.

2.10 Data Collection

The data we use for our work come from the World Health Organization (WHO) [3] website which contains information on Ebola registrations covering the year 2014 to 2016 in various countries from which we have extracted data for Guinea, namely:

- Total Cases of infected people from the first day of registration to time t;
- Total Deaths the total number of deaths during the epidemic period.

We then use the Euler discretizations of the SEIRD-B model to isolate the dynamic parameters, namely $\beta_1(t)$, $\beta_2(t)$, $d(t)$ and $\tau(t)$. The effective reproduction number is then calculated. These parameters are then estimated with LSTM and GRU and used to discretize the model to obtain $\hat{I} and \hat{D}$ with the LSTM and GRU parameters, which will be compared with the data from Ebola Guinea in the database (Fig. 6).

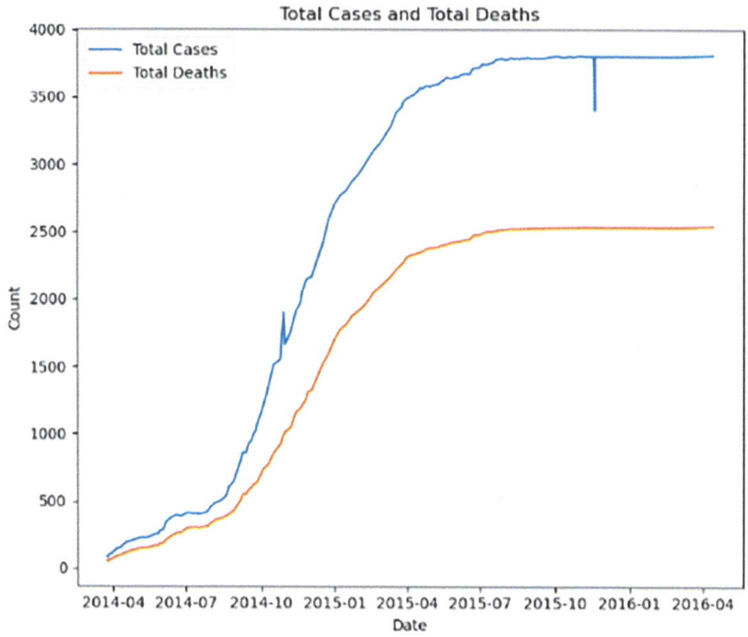

Fig. 6 Number of infected and deaths in Guinea between April 2014 and April 2016

2.11 Data Preprocessing

A crucial step in data analysis is data preprocessing, which is typically used to convert the original data into a format that can be easily modified to enable the best possible data analysis process. The original data may contain inconsistencies, such as noisy or missing values, which is why data preprocessing is necessary. Data preprocessing can be done in accordance with different goals that are specific to the use of the algorithms. Preprocessing generally consists of:

- Handling missing values: this will involve either deleting rows/columns that contain missing values, or replacing missing values using either the median, mean, or mode. Imputation by interpolation can be used for temporal data;
- Remove duplicates: to avoid bias, identify duplicate lines and remove them from the data;
- Outlier management: use statistical methods such as inter-quartile distances to detect outliers and remove them if necessary, or reduce the impact of outliers by applying logical or logarithmic transformations [42, 43].

2.12 Standardization (Z-Score)

A data preprocessing method called normalization is used to bring quantitative variables into line with one another. Normalization's primary goal is to remove the bias caused by disparate measurement scales between variables by transforming the data into a predetermined range of values. Normalizing features has two benefits: it makes gradient descent-based machine learning models easier to implement; it equalizes the weight of each dimension to prevent variables with high values from being regarded as more significant than those with low values; and it standardizes the mean of the data's standard deviation to make the learning problem simpler. The standardization technique (Z-score) is based on using the mean and standard deviation of the variables to center and reduce them. It is defined by the formula:

$$X_{\text{stand}} = \frac{X - \mu}{\sigma}, \quad (12)$$

with:

- μ: the average of the data;
- σ: the standard deviation.

In order to increase the convergence performance of learning algorithms, normalizing approaches are useful for distance-based algorithms like K-means and linear regression. This technique is required when the data must be centered around 0 with a standard deviation of 1.

The number of people infected and deceased in Guinea is not normally distributed. The visualization allows us to see that the evolution of these two cases is asymmetrical and that the curves are distributed more positively to the right (see Fig. 7). The Z-score is used to normalize the data so that it is centered on 0 with a standard deviation of 1, in order to avoid overfitting during training. The choice to use Z-scores is based on the adaptability of the tanh activation function in the internal layers for both algorithms, LSTM and GRU, which stabilizes the gradients and improves convergence for predictions, as shown by the results. The use of Z-scores in our work serves for normalization, with the aim of bringing the data to the same scale in order to train it with neural networks, and not for testing probability hypotheses or classical predictions, for which we do not need to calculate the confidence interval.

2.13 Training Set (80%)

Eighty percent of the entire data set is the training set. With the use of this set, the model is trained, or its parameters are changed to reduce errors and increase prediction accuracy. In this stage, the model gains knowledge of the connections and trends found in the given data. Optimizing the model's parameters to enable precise predictions on fresh data is the goal.

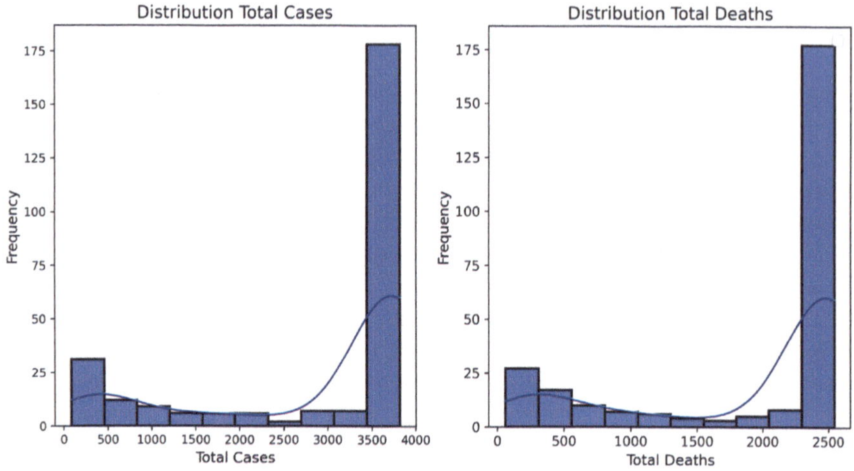

Fig. 7 Distribution curve of infected cases and deaths

2.14 Validation Set (20%)

The remaining 20% of the data is represented by the test set. After training, the model's performance is assessed using this set. Checking the model's performance on data it hasn't encountered during training is the primary goal. To make sure the model is not just over-fitted to the training data, but also has good generalization, this assessment is essential. We can get a more accurate gauge of the model's capacity to forecast fresh data by testing it on this distinct data set.

2.15 Theoretical Calculation of Some Model Parameters

The four model parameters that we will theoretically determine are the lethality rate ($d(t)$), the rate of natural decomposition of the virus in the environment ($\tau(t)$), the rate of transmission of pathogens from the environment to susceptible ($\beta_1(t)$), and the rate of transmission of pathogens from deceased bodies to susceptible ($\beta_2(t)$). The other parameters will stay constant. To do this, we isolate the parameters to be found using Euler discretizations after using the differential equation representing the $SEIRD - B$ model.

$$
\begin{cases}
\dfrac{dS(t)}{dt} = \Delta - [\beta_1(t)B(t)(1+\alpha_1 B(t)) + \beta_1(t)\psi_1 E(t)(1+\alpha_2 E(t)) \\
\qquad +\beta_1(t)\psi_2 I(t)(1+\alpha_3 I(t))]\, S(t) - \beta_2(t)S(t)D(t)(1+\alpha_4 D(t)) - \mu S(t), \\
\dfrac{dE(t)}{dt} = [\beta_1(t)B(t)(1+\alpha_1 B(t)) + \beta_1(t)\psi_1 E(t)(1+\alpha_2 E(t)) \\
\qquad +\beta_1(t)\psi_2 I(t)(1+\alpha_3 I(t))]\, S(t) \\
\qquad + \beta_2(t)S(t)D(t)(1+\alpha_4 D(t)) - (b_1+\delta_1+\mu)E(t), \\
\dfrac{dI(t)}{dt} = b_1 E(t) - (\delta_2 + d(t) + \mu)I(t), \\
\dfrac{dR(t)}{dt} = \delta_1 E(t) + \delta_2 I(t) - \mu R(t), \\
\dfrac{dD(t)}{dt} = d(t)I(t) - \phi D(t), \\
\dfrac{dB(t)}{dt} = \gamma_1 E(t) + \gamma_2 I(t) + \gamma_3 D(t) - \tau(t)B(t).
\end{cases}
\tag{13}
$$

Let $N(t) = S(t) + E(t) + I(t) + R(t)$ the total population of the geographical area (Guinea). The effective reproduction number $(R_0(t))$ (see [19] for how it was obtained) will be determined as a function of the parameters $\beta_1(t)$, $\beta_2(t)$, $d(t)$, and $\tau(t)$, by

$$R_0(t) = R_E(t) + R_I(t) + R_D(t) + R_B(t). \tag{14}$$

with

$$
\begin{aligned}
R_E(t) &= \frac{\Delta\, \psi_1\, \beta_1(t)}{(b_1+\delta_1+\mu)S(t)}, \\
R_I(t) &= \frac{\Delta\, \psi_2\, \beta_1(t)\, b_1}{(b_1+\delta_1+\mu)(\delta_2+d(t)+\mu)S(t)}, \\
R_D(t) &= \frac{\Delta\, \beta_2(t)\, d(t)}{(b_1+\delta_1+\mu)(\delta_2+d(t)+\mu)\phi}, \\
R_B(t) &= \frac{\Delta\, [\beta_1(t)\gamma_3 d(t) + (\gamma_1(\delta_2+d(t)+\mu)+\gamma_2 b_1)\beta_1(t)\phi]}{\mu(b_1+\delta_1+\mu)(\delta_2+d(t)+\mu)\tau(t)\phi}
\end{aligned}
$$

We assume that the total population of Guinea in 2014 is 10,523,361 and that life expectancy is 57.89 years. The natural mortality rate μ being equal to $\dfrac{1}{57.89 \times 365}$.

Using Euler's method (see [44] for further explanation) to discretise the ordinary differential equations (ODEs) of the SEIRD-B model, and taking the discretization step equal to one day, the model 13 can be written in the following form:

$$\begin{cases} S(t+1) = S(t) + [\Delta - (\beta_1(t)B(t)(1+\alpha_1 B(t)) + \beta_1(t)\psi_1 E(t)(1+\alpha_2 E(t)))\, S(t) \\ \qquad -\beta_1(t)\psi_2 I(t)(1+\alpha_3 I(t))S(t) - \beta_2(t)S(t)D(t)(1+\alpha_4 D(t)) - \mu S(t)], \\ E(t+1) = E(t) + [(\beta_1(t)B(t)(1+\alpha_1 B(t)) + \beta_1(t)\psi_1 E(t)(1+\alpha_2 E(t)))\, S(t) \\ \qquad +\beta_1(t)\psi_2 I(t)(1+\alpha_3 I(t))S(t) + \beta_2(t)S(t)D(t)(1+\alpha_4 D(t))] \\ \qquad - (b_1(t) + \delta_1(t) + \mu)E(t), \\ I(t+1) = I(t) + [b_1 E(t) - (\delta_2 + d(t) + \mu)I(t)], \\ R(t+1) = R(t) + [\delta_1 E(t) + \delta_2 I(t) - \mu R(t)], \\ D(t+1) = D(t) + [d(t)I(t) - \phi D(t)], \\ B(t+1) = B(t) + [\gamma_1 E(t) + \gamma_2 I(t) + \gamma_3 D(t) - \tau(t)B(t)]. \end{cases} \tag{15}$$

As we have the data on infected, dead, and the knowledge of constant parameters, we can therefore determine the theoretical solutions of the short time dependent parameters namely: $\beta_1(t)$, $\beta_2(t)$, $d(t)$, and $\tau(t)$ by solving the systems of equations. This solution is obtained after applying Algorithm 1.

The calculation of terms and the effective reproduction number is obtained with Algorithm 2

2.16 Prediction of the Dynamic Parameters of SEIRD-B with LSTM and GRU

Following the acquisition of the theoretical parameters outlined in the aforementioned algorithm, we separate them into 80% training and 20% testing. The LSTM and GRU models then use the training data as input, and predictions are made using the 20% test data. To improve accuracy, we employ two hidden layers for the LSTM and GRU techniques, and we train them using the Adam optimizer. Table 1 displays the remaining parameters, including the batch size, the number of hidden units in each layer, and the epoch numbers. Algorithm 3 shows how time-varying parameter learning and prediction using LSTM/GRU are done.

2.17 Simulation of the Infected and Deceased

After estimating the dynamic parameters, we apply them in Algorithm 4 for the discrete SEIRD-B model to predict the cases of the Infected and the Dead.

Algorithm 1: Part 1: Initialization and calculation of basic parameters.

Data: Epidemiological data on Ebola (`Ebola_Data.xlsx`)
Result: Estimation of model parameters β_1, β_2, ϕ, d, and R_{eff}
Input : Time data t, Total cases I, Deaths D.
Output: List S, E, I, D, R, B, β_1, β_2, d, ϕ, R_{eff}

Initialization of the total population $N_0 = 11.33 \times 10^6$;
Initialization of model parameters: μ, Δ, ϕ, β_1, β_2, γ_1, γ_2, etc.;
Compute E_0 et S_0 based on available data;
Initialization of $B_0 = 100000000$;
Add E_0, S_0, and B_0 to the corresponding lists;
for $t = 0$ **to** *length*$(I) - 1$ **do**
    `/* Computation of the different variables and parameters for each day */`
    `/* Computation of` $d(t)$ `: */`
    Compute $dD/dt = D[t+1] - D[t]$;
    Compute $d(t) = \frac{dD/dt + \phi D[t]}{I[t]}$;
    `/* Computation de` $E(t)$ `: */`
    Computation $dI/dt = I[t+1] - I[t]$;
    Computation $E(t) = \frac{dI/dt + (\delta_2 + d(t) + \mu) I[t]}{b_1}$;
    `/* Computation of` $R(t)$ `: */`
    Compute $dR/dt = \delta_1 E[t] + \delta_2 I[t] - \mu R[t]$;
    Update $R[t+1] = R[t] + dR/dt$;
    `/* Computation of` $B(t)$ `: */`
    Compute $dB/dt = \gamma_1 E[t] + \gamma_2 I[t] + \gamma_3 D[t] - \tau B[t-1]$;
    Update $B[t+1] = B[t] + dB/dt$;
    `/* Computation of` $\tau(t)$ `: */`
    Compute $\tau(t) = \frac{\gamma_1 E[t] + \gamma_2 I[t] + \gamma_3 D[t]}{B[t]}$;
    `/* Computation of` $S(t)$ `: */`
    Compute $S[t+1] = N_0 - E[t] - R[t] - I[t]$;
    Add $\beta_1[t]$, $\beta_2[t]$, $d[t]$, $\phi[t]$ to the corresponding lists;
end

3 Results

3.1 Evaluation Metrics

We use three metrics to evaluate the performance of the results, the root mean square error (RMSE), mean absolute percentage error (MAPE), the coefficient of determination R^2. Let N be the number of data points, y_i the true value of the i-th data point, and $\hat{y}_i$ the prediction of y_i.

Algorithm 2: Part 2: Calculation of terms and the effective reproduction number.

Data: Epidemiological data on Ebola (`Ebola_Data.xlsx`)
Result: Estimation of model parameters β_1, β_2, ϕ, d, and R_{eff}
Input : Time data t, Total cases I, Deaths D
Output: List R_{eff}, R_E, R_I, R_D, R_B

for $t = 0$ **to** $length(I) - 2$ **do**
    /* Computation of $R_{\text{eff}}(t)$ */
    /* Computation of $R_E(t)$: */
    Compute $R_E(t) = \frac{\Delta \psi_1 \beta_1[t]}{(b_1+\delta_1+\mu)S[t]}$;
    /* Computation of $R_I(t)$: */
    Compute $R_I(t) = \frac{\Delta \psi_2 \beta_1[t] b_1}{(b_1+\delta_1+\mu)(\delta_2+d(t)+\mu)S[t]}$;
    /* Computation of $R_D(t)$: */
    Compute $R_D(t) = \frac{\Delta \beta_2[t] d(t)}{(b_1+\delta_1+\mu)(\delta_2+d(t)+\mu)\phi}$;
    /* Computation of $R_B(t)$: */
    Compute $R_B(t) = \frac{\Delta[\beta_1[t]\gamma_3 d(t)+(\gamma_1(\delta_2+d(t)+\mu)+\gamma_2 b_1)\beta_1[t]\phi]}{\mu(b_1+\delta_1+\mu)(\delta_2+d(t)+\mu)\tau(t)\phi}$;
    /* Computation of $R_{\text{eff}}(t)$: */
    Compute $R_{\text{eff}}(t) = R_E(t) + R_I(t) + R_D(t) + R_B(t)$;
    Add $R_E[t]$, $R_I[t]$, $R_D[t]$, $R_B[t]$, and $R_{\text{eff}}(t)$ to the corresponding lists;
end

Table 1 Summary of hyperparameters for the LSTM and GRU models

Hyperparameter	Value
Number of hidden layers	2
LSTM/GRU units (first layer)	100
LSTM/GRU units (second layer)	50
Activation function	tanh
Optimizer	Adam
Learning rate	0.001 (per default)
Number of epochs	200
Batch size	32
Sliding window (time step)	3

3.1.1 Root Mean Squared Error (RMSE)

RMSE is the square root of the MSE. It is expressed in the same unit as the target variable, which makes it more interpretable. The RMSE retains the property of penalizing large errors more heavily.

Algorithm 3: Time-varying parameter learning and prediction using LSTM/GRU.

Input: *data* (Numerical data of $\beta_1(t), \beta_2(t), d(t), \tau(t), R_{\text{effective}}(t)$)
Input: epochs, timesteps, batch_size
Output: Prediction and parameter adjustment $\beta_1(t), \beta_2(t), d(t), \tau(t), R_{\text{effective}}(t)$

```
Function Standardize(data):
    sc = StandardScaler();
    data = sc.fit_transform(np.float64(data));
    return sc, data;
Function SplitData(data, a = 0.8):
    l ← length(data);
    train ← data[0 : l * a];
    test ← data[l * a : l];
    return train, test;
Function CreateDataset(train, timesteps):
    x_train ← [];
    y_train ← [];
    for i ← timesteps to l * a + timesteps do
        x_train.append(train[i − timesteps : i, 0]);
        y_train.append(train[i : i + timesteps, 0]);
    end
    return x_train, y_train;
Function FitLSTM(x_train, y_train, batch_size, timesteps, neurons):
    model ← Sequential();
    if model_type == "LSTM" then
        model.add(LSTM(neurons, stateful = True));
        model.compile(optimizer =' adam', loss =' mse');
    end
    else if model_type == "GRU" then
        model.add(GRU(neurons, stateful = True));
        model.compile(optimizer =' adam', loss =' mse');
    end
    for i ← 1 to epochs do
        model.fit(x_train, y_train, shuffle = False, epochs = 1, batch_size =
          batch_size);
        model.reset_states();
    end
    return model;
Function Predict(model, x_train):
    y_predict ← model.predict(x_train);
    return y_predict;
Function InverseTransform(sc, y_predict):
    y_predict ← sc.inverse_transform(y_predict);
    return y_predict;
Function PlotResults(test, y_predict):
    plt.plot(test);
    plt.plot(y_predict);
    plt.show();
```

Algorithm 4: Simulation of SEIRDB compartments with LSTM/GRU time parameters.

Input: N: Total population of the study area
Input: T: Number of days or iterations of the simulation
Input: $S_0, E_0, I_0, R_0, D_0, B_0$: Initial values for susceptible, exposed, infected, withdrawn, deceased and non-diseased compartments
Input: $\beta_1(t), \beta_2(t), d(t), \tau(t)$: Time series of transmission, death and response parameters for non-diseased people
Input: $\Delta, \alpha_1, \alpha_2, \alpha_3, \alpha_4, \psi_1, \psi_2, b_1, \delta_1, \delta_2, \mu, \phi, \gamma_1, \gamma_2, \gamma_3$: SEIRD-B model parameters
Output: S, E, I, R, D, B: Time series of compartments after simulation
Output: Parameter predicted $\hat{\beta}_1, \hat{\beta}_2, \hat{d}, \hat{\tau}$

for $t \leftarrow 0$ *to* $T - 1$ **do**
  /* Update of parameters $\beta_1(t), \beta_2(t), d(t), \tau(t)$ with the values predicted by the LSTM/GRU model */
  $\beta_1(t) \leftarrow \hat{\beta}_1(t)$;
  $\beta_2(t) \leftarrow \hat{\beta}_2(t)$;
  $d(t) \leftarrow \hat{d}(t)$;
  $\tau(t) \leftarrow \hat{\tau}(t)$;
  /* SEIRD-B simulation using Euler's method for the values of each compartment */
  $S_{t+1} \leftarrow S_t + \Delta - \beta_1(t) \cdot B_t \cdot (1 + \alpha_1 \cdot B_t) + \beta_1(t) \cdot \psi_1 \cdot E_t \cdot (1 + \alpha_2 \cdot E_t) + \beta_1(t) \cdot \psi_2 \cdot I_t \cdot (1 + \alpha_3 \cdot I_t)$;
  $E_{t+1} \leftarrow E_t + [\beta_1(t) \cdot B_t \cdot (1 + \alpha_1 \cdot B_t) + \beta_1(t) \cdot \psi_1 \cdot E_t \cdot (1 + \alpha_2 \cdot E_t) + \beta_1(t) \cdot \psi_2 \cdot I_t \cdot (1 + \alpha_3 \cdot I_t)]N - (b_1 + \delta_1 + \mu) \cdot E_t$;
  $I_{t+1} \leftarrow I_t + b_1 \cdot E_t - (\delta_2 + d(t) + \mu) \cdot I_t$;
  $R_{t+1} \leftarrow R_t + \delta_1 \cdot E_t + \delta_2 \cdot I_t - \mu \cdot R_t$;
  $D_{t+1} \leftarrow D_t + d(t) \cdot I_t - \phi \cdot D_t$;
  $B_{t+1} \leftarrow B_t + \gamma_1 \cdot E_t + \gamma_2 \cdot I_t + \gamma_3 \cdot D_t - \tau(t) \cdot B_t$;
end

$$\text{RMSE} = \sqrt{\text{MSE}} = \sqrt{\frac{1}{n}\sum_{i=1}^{n}(\hat{y}_i - y_i)^2}. \tag{16}$$

RMSE gives a more intuitive estimate of the standard deviation of the prediction errors. As with MSE, a lower value of RMSE indicates a more accurate model. It is often used as the main metric due to its direct interpretation [45].

3.1.2 Measures the Error in Percentage (MAPE)

The percentage error is a measure of the difference between the actual values and the predicted values, which is estimated as a percentage. This measure is used to assess the accuracy of an estimate. The MAPE is calculated using the following formula:

$$\text{MAPE} = \frac{1}{n}\sum_{i=1}^{n}\frac{|y_i - \hat{y}_i|}{y_i} \times 100. \quad (17)$$

3.1.3 Coefficient of Determination (R^2)

The coefficient of determination R^2 measures the proportion of the total variance in the data explained by the model. This is a unitless measure that varies between 0 and 1. A value close to 1 indicates that the model explains a large proportion of the variance in the data [46].

$$R^2 = 1 - \frac{\sum_{i=1}^{n}(y_i - \hat{y}_i)^2}{\sum_{i=1}^{n}(y_i - \bar{y})^2}, \quad (18)$$

where $\bar{y}$ is the average of the real values. We have the following cases:

- $R^2 = 1$ means that the model perfectly explains all the variations in the data;
- $R^2 = 0$ indicates that the model does not explain any of the variations in the data;
- A negative value of R^2 can occur if the model is worse than a regression model that simply predicts the mean of the actual values.

3.2 *Parameter Estimations*

3.2.1 Transmission Rates: β_1 and β_2

Figure 8 depicts the variation the transmission rates β_1 and β_2.

3.2.2 Disease Induced Death d and Decay Decomposition Rate τ

Figure 9 depicts the variation of lethality rate d and virus decomposition rate in the environment τ between 2014 and 2016 obtained with the LSTM and GRU networks.

3.2.3 Effective Reproduction Number $\mathcal{R}_0(t)$

Figure 10 presents the variation in the effective reproduction rate $\mathcal{R}_0(t)$ between 2014 and 2016 obtained with the LSTM and GRU networks.

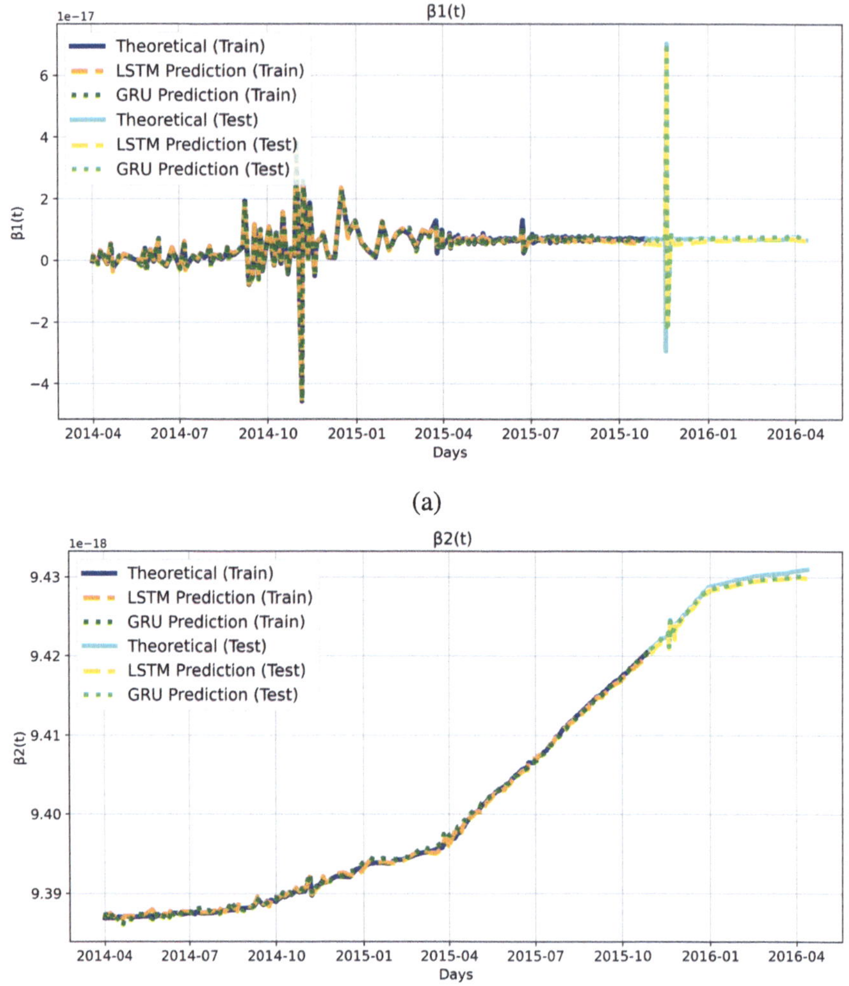

(a)

(b)

Fig. 8 Variation in transmission rate between 2014 and 2016 obtained with the LSTM and GRU networks: **a** for β_1 and **b** for β_2. The blue curves represent the theoretical values of β_1 and β_2 for the training data set and the cyan (dotted) curve for the test data set, for both transmission rates, the orange (dotted) curve shows the predictions obtained with the LSTM network on the training data set and the orange curve for the test data set. For the predictions obtained with the GRU network, the green (dotted) curve shows the predictions on the training set and Orange on the test set

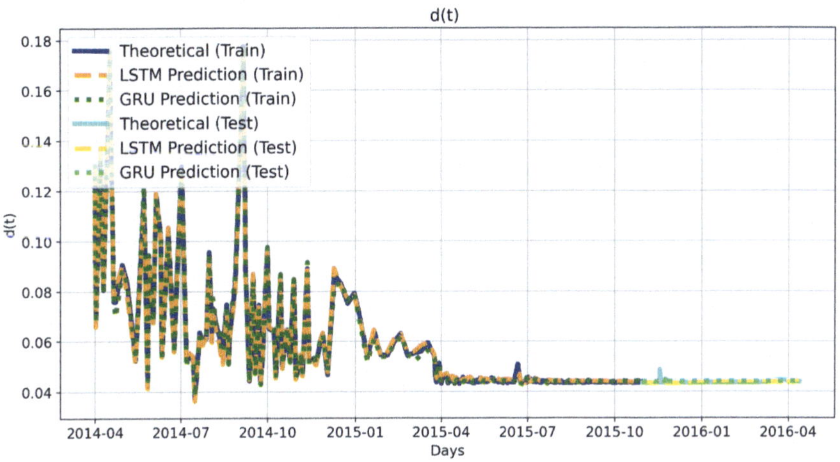

(a)

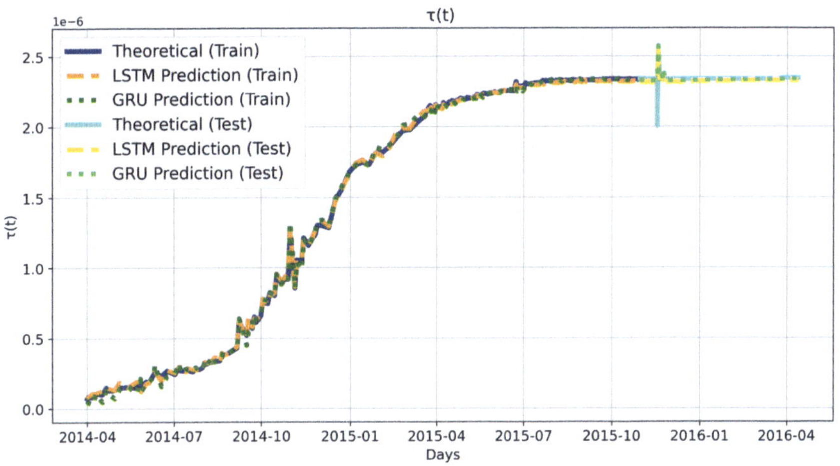

(b)

Fig. 9 Variation in lethality rate d and virus decomposition rate in the environment τ between 2014 and 2016 obtained with the LSTM and GRU networks. (The blue curves represent the theoretical values for the training data set and the cyan curve (dotted line) for the test set. The orange (dotted) curve indicates the predictions obtained with the LSTM network on the training set and orange for the test data set. For the predictions obtained with the GRU network, the green (dotted) curve shows the predictions on the training set and Orange on the test set

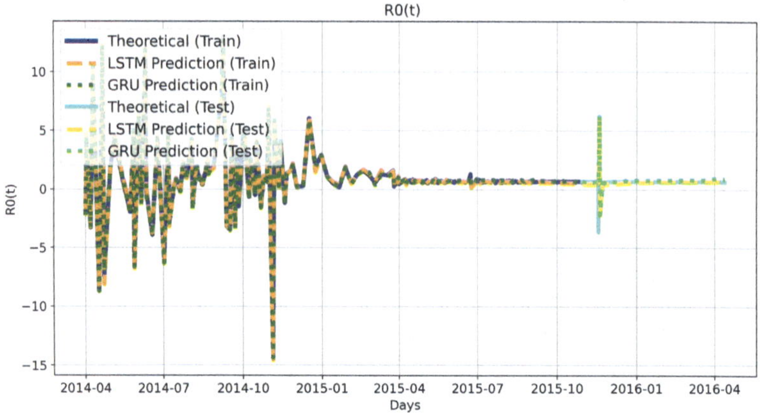

Fig. 10 Variation in effective reproduction rate $\mathcal{R}_0(t)$ between 2014 and 2016 obtained with the LSTM and GRU networks. The blue curve represents the theoretical values for the training data set and the cyan (dotted) curve for the test data set. The orange (dotted) curve shows the predictions obtained with the LSTM network on the training set and orange for the test data set. For the predictions obtained with the GRU network, the green (dotted) curve shows the predictions on the training set and orange on the test set

3.2.4 Mean and Variance of Estimated Parameters

Table 2 presents the mean and variance of each estimated model parameter. From Table 2, we observe that the variances and means of the predicted parameters are closer to the theoretical parameters. The LSTM model seems slightly more accurate in terms of mean and variance for certain parameters, while the GRU offers better stability in predictions.

3.2.5 Discretization of the SEIRD-B Model with Predicted Parameters

Figure 11 shows the prediction of total cases and total deaths using both methods.

3.2.6 SEIRD-B LSTM and SEIRD-B GRU Error Results

Absolute error of total Cases (a), relative error of total Cases (c). and absolute error of total deaths (b), (d) for the relative error with the parameters predicted with LSTM and GRU (SEIRD-B LSTM and SEIRD-B GRU), are depicted in Fig. 12.

Table 2 Mean and variance of the 3 methods

Parameter	Type of data	Mean	Variance
$\beta_1(t)$	Theoretical	5.734205×10^{-18}	5.191103×10^{-35}
$\beta_1(t)$	LSTM prediction	5.755055×10^{-18}	4.673730×10^{-35}
$\beta_1(t)$	GRU prediction	5.716582×10^{-18}	4.680075×10^{-35}
$\beta_2(t)$	Theoretical	9.405709×10^{-18}	1.968803×10^{-40}
$\beta_2(t)$	LSTM prediction	9.405935×10^{-18}	1.968810×10^{-40}
$\beta_2(t)$	GRU prediction	9.406161×10^{-18}	1.970590×10^{-40}
$d(t)$	Theoretical	5.284876×10^{-2}	4.049441×10^{-4}
$d(t)$	LSTM prediction	5.248558×10^{-2}	3.917995×10^{-4}
$d(t)$	GRU prediction	5.377638×10^{-2}	4.142663×10^{-4}
$\tau(t)$	Theoretical	1.794337×10^{-6}	6.572213×10^{-13}
$\tau(t)$	LSTM prediction	1.814325×10^{-6}	6.390290×10^{-13}
$\tau(t)$	GRU prediction	1.816459×10^{-6}	6.357411×10^{-13}
$R_0(t)$	Theoretical	9.016904×10^{-1}	6.485874×10^{0}
$R_0(t)$	LSTM prediction	8.548062×10^{-1}	5.833552×10^{0}
$R_0(t)$	GRU prediction	9.102941×10^{-1}	5.672642×10^{0}

3.2.7 Results of Total Cases and Total Deaths Training with LTSM and GRU

Figure 13 shows the results of the Ebola data trained with the two deep learning tools (LSTM and GRU). Table 3 presents the comparison of both methods using two different metrics, RMSE, MAPE, and R^2.

4 Discussion

Figure 8 represents the prediction of $\beta_1(t)$ and $\beta_2(t)$ using LSTM and GRU. These figures show the variations of the two transmission rates during the Ebola epidemic in 2014 to 2016, we interpret these figures as follows:

- $\beta_1(t)$: We visualize a high variation of $\beta_1(t)$ towards the month of August and September 2014 but, a considerable drop at the beginning of November of 2014 and at the beginning of January 2015, there is again the high rate and remains almost constant with slight variations from April 2014 until October of which we notice again the peak of all the period and the rate remains constant from January 2016 to April at an average of 4.10–35.
- $\beta_2(t)$: We see that the rate increases from the beginning in March 2014 until the end of the recording period in April 2016.

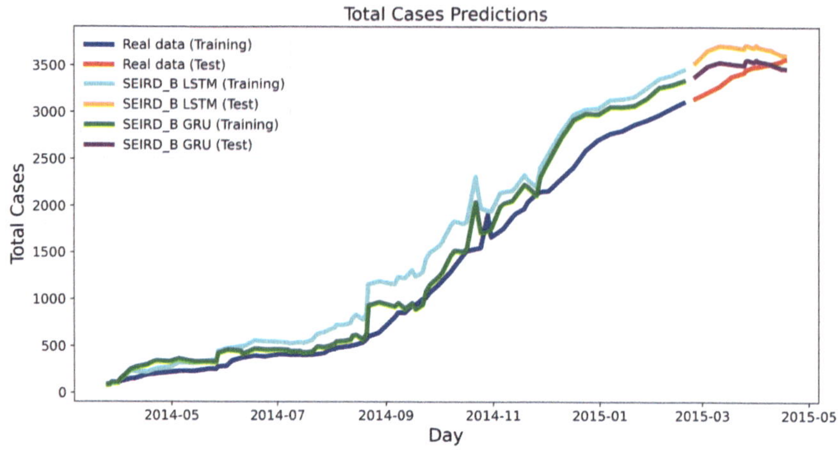

(a)

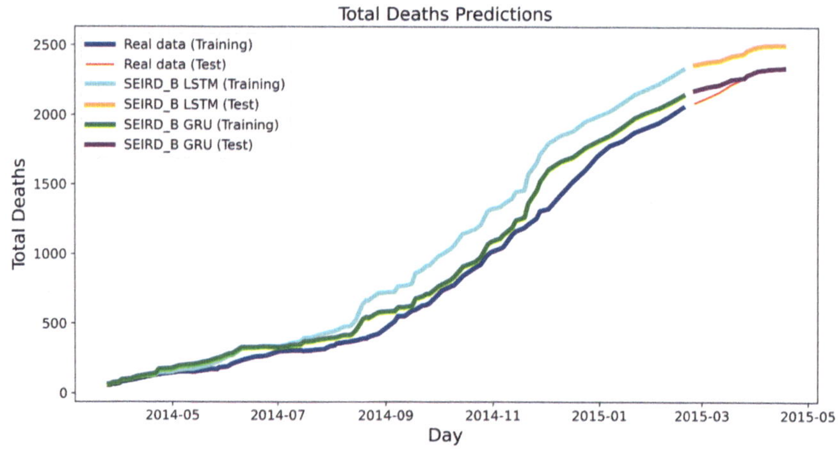

(b)

Fig. 11 Simulation of Total Cases (**a**) and Total Deaths (**b**) with parameters predicted with LSTM and GRU (SEIRD-B LSTM and SEIRD-B GRU)

Figure 9 is the prediction values of d and τ during the epidemic registration period from 2014 to 2016 with LSTM and GRU.

- d: The mortality rate due to disease d, is interpreted as follows: From the start of the epidemic we can see that the rate is high in April 2014 at 0.8 from where a drop in May 2014 to 0.04 and June again a rise in September 2014 and slight fluctuations until April 2015 which remains constant until 2016 at the last recording date.
- τ: The rate of virus degradation in the environment starts to increase from April 2014 with peaks and starts to be stable from April 2015 until April 2016.

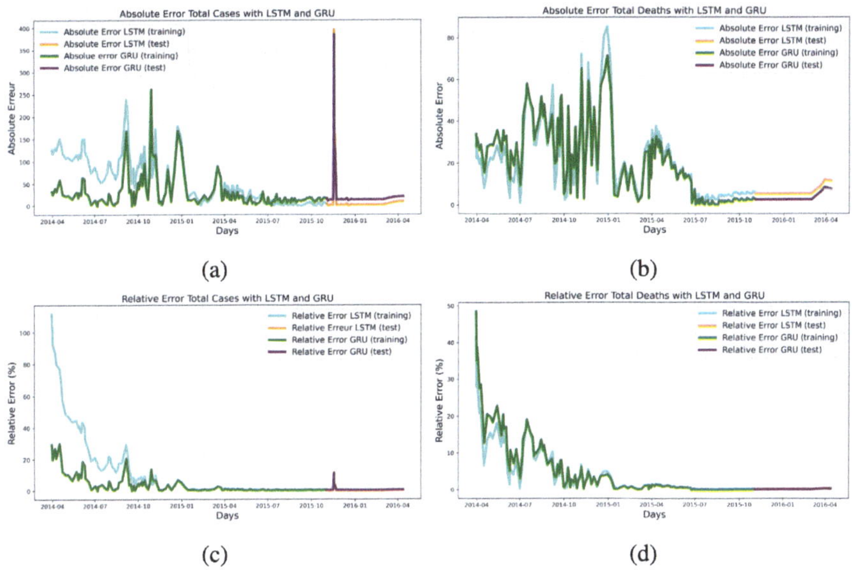

Fig. 12 Absolute error of total Cases (**a**), relative error of total Cases (**c**) and absolute error of total deaths (**b**), (**d**) for the relative error with the parameters predicted with LSTM and GRU (SEIRD-B LSTM and SEIRD-B GRU)

Table 3 Validation of metrics for the LSTM, GRU, SEIRD-B LSTM and SEIRD-B GRU models

State variables	Methods	RMSE	MAPE (%)	R^2
$I(t)$	LSTM	56.729	2.01	0.9987
	GRU	52.187	4.46	0.9994
	SEIRD-B LSTM	184.718	10.89	0.9784
	SEIRD-B GRU	142.044	13.02	0.9872
$D(t)$	LSTM	21.619	1.72	0.9994
	GRU	18.506	1.79	0.9995
	SEIRD-B LSTM	96.483	8.50	0.9862
	SEIRD-B GRU	83.770	11.30	0.9896

Figure 10 is the prediction of the number of effective reproduction R0(t) between 2014 and 2016 obtained with LSTM and GRU, we note that in April 2014 an infected person can transmit up to 10 people in an uninfected population, which explains the outbreak of the epidemic in 2014. A decrease in R0(t) will be noticed towards the end of April but again an outbreak or peak of 10 June and September 2014 and a considerable decrease at the end of October 2014, which again increases in November 2015, where an infected person can transmit to more than 5 people the epidemic and which will decrease by more than 2 people in April 2015 until April 2016.

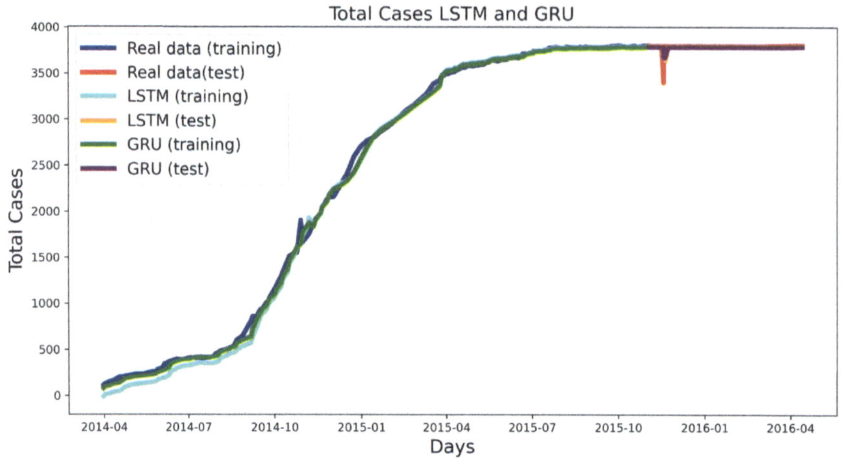

(a)

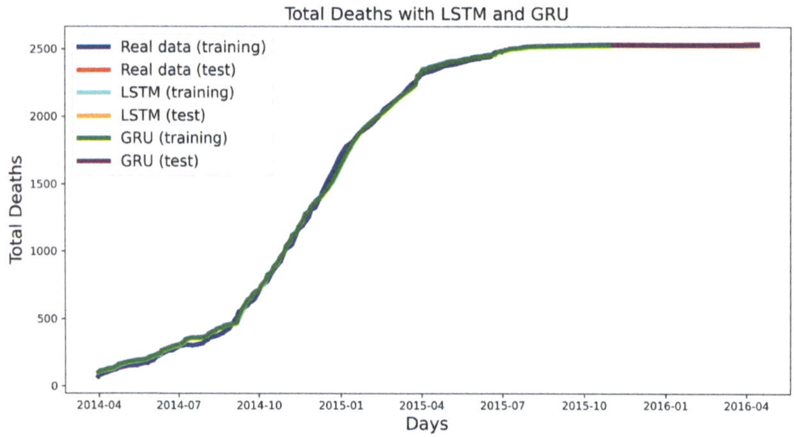

(b)

Fig. 13 Forecast of total number of cases (**a**) and total number of deaths (**b**) using LSTM and GRU

Unlike previous study [19], the values of β_1 and β_2 depend on time and therefore reflect the control measures taken by the Guinean authorities to reduce the number of cases and deaths. Also, according to Fig. 10, the maximum value of the basic reproduction number $\mathcal{R}_0$ is less than 10, which contradicts the value obtained in [19], which was 37. This shows that the combination of the deterministic model and the LSTM and GRU methods allows for a better assessment of the quantitative dynamics of the disease and the effectiveness of the control measures put in place.

To illustrate the advantage of our proposed model, we compare it with the LSTM and GRU models directly to the data. We thus have four methods LSTM, GRU, SEIRD-B LSTM and SEIRD-B GRU. Figure 11 is the result of Total Cases (I) and Total Deaths (D) with the SEIRD-B method with the parameters estimated with LSTM (SEIRD-B LSTM) and the parameters with GRU (SEIRD-B GRU) and Fig. 12 is the absolute error and the relative error of I and D. Figure 13 is the prediction of I and D using the direct real data with the two neural network algorithms LSTM and GRU. Based on the results, GRU and SEIRD-B GRU methods are better adopted to Guinea Ebola data from 2014 to 2016, Fig. 11 verifies and explains well by the result of absolute and relative errors of Fig. 12 which can be seen that GRU relative and absolute errors are lower and stable and close to 0 with those of LSTM. In addition, we evaluated the performance of the four methods by calculating RMSE, MAPE and R^2 shown in Table 3. The GRU model for I has RMSE = 52.187, MAPE = 4.46% and $R^2 = 0.9994$ and SEIRD-B GRU has the metrics RMSE = 142.044, MAPE = 13.02% and $R^2 = 0.9872$. The D metrics with GRU have RMSE = 18.506, MAPE = 1.79% and $R^2 = 0.9995$ and SEIRD-B GRU have RMSE = 83.770, MAPE = 11.30% and $R^2 = 0.9896$ which are performing results with low RMSE and R^2 closer to 1.

We have used the Adam optimizer for training with LSTM and GRU in order to adapt the learning rate. In order to find an optimal solution, we carried out several simulations and one of the limitations is the lack of data for the test, which is 20% of 263 records.

The main goal of employing Gated Recurrent Units (GRU) and Long Short-Term Memory (LSTM) is to get around the drawbacks of conventional Recurrent Neural Networks (RNNs), particularly the vanishing gradient issue, which prevents RNNs from learning long-term dependencies in sequential data. In order to handle complex temporal relationships more effectively in tasks like time series prediction and natural language processing, LSTMs and GRUs introduce gating mechanisms that regulate the flow of information. This enables them to selectively remember or forget information over extended sequences. In the same way that past experiences are used to make sense of the present, RNNs allow information from previous points in a sequence to be physically stored thanks to their memory or hidden state.

5 Conclusion

This study focused on predicting new cases of infection and deaths during the Ebola epidemic in Guinea. It was based on an innovative methodology, combining the SEIRD-B epidemiological model and deep learning techniques, in particular the LSTM and GRU recurrent neural networks. The main objective was to estimate the fundamental dynamic parameters in order to model the spread of the epidemic at a high case-fatality rate and make predictions.

The methodological approach followed is structured in several key steps. Firstly, the SEIRD-B study model was formulated in mathematical form. This model, in which the transition from one compartment to another depends on parameters, was then discretized to enable numerical simulation. The dynamic parameters were first obtained by solving the model equations and then trained with LSTM and GRU, which are particularly well suited to complex non-linear time series. The parameters estimated using these algorithms were used in the numerical simulation. In comparison, the 2 deep learning algorithms were trained on the Ebola epidemiological dataset, in particular total cases and deaths, and then validated by comparing predictions with empirical observations. The performance of the models was evaluated using metrics such as RMSE, MAPE and R^2.

The results highlight the relevance of recurrent neural networks for predicting epidemic dynamics. The SEIRD-B GRU and GRU models were found to be slightly superior, providing greater stability and good accuracy of estimates. The predictions from the models showed a strong correlation with the real data, confirming their ability to capture the complex variations in epidemiological parameters. Nevertheless, the study also showed some limitations, the main one being the instability of the gradient encountered when training the networks, which can be explained by the complexity of the model and the noisy nature of the input data. In addition, dependence on available epidemiological data poses a challenge, as variations in data quality and frequency can alter the accuracy of predictions. These limitations call for improvements in future methodologies. Possible improvements include integrating hybrid methods combining ensemble methods to estimate parameters, or combining least squares techniques and epidemiological deep learning methods to improve the robustness of estimates. To explore an extension of the SEIRD-B model by adding the vaccinated compartment in line with epidemiological reality.

The main advantage of this approach is that the SEIRD-B hybrid approach with estimated parameters effectively captures non-linear dynamics and adapts well to temporal variations in epidemiological data. Nevertheless, considerable computational complexity, and performance depends heavily on the quality and, above all, the quantity of training data.

This study is important for public health, where accurate predictions can guide real-time interventions. The integration of the deterministic SEIRD-B compartmental model with neural networks (LSTM, GRU) to estimate certain model parameters is an approach that overcomes the limitations of statistical methods with non-linear data and variables. In summary, this study demonstrates the potential of the LSTM and GRU approaches for predicting dynamics. It represents a step forward in the planning of health crisis interventions and therefore opens up prospects for the integration of AI techniques in the management of complex epidemiological crises.

As a future prospect, we suggest applying other neural network methods for parameter estimation, including ensemble models combining LSTM and GRU, as well as hybrid models to improve performance. We also suggest applying these models studied for Ebola in Guinea to other countries and other epidemics. In this work, ordinary differential equation with integer-order derivative is used to formulate the model. Fractional-order derivatives will be explored as future avenues of research.

Competing Interests The authors have no conflicts of interest to declare that are relevant to the content of this chapter.

Ethics Approval No applicable.

Acknowledgements The authors thank the `King of the Skolabwandjan` for his multi-faceted assistance which made possible to complete this work.

References

1. Comité régional de l'Afrique. Épidémie de maladie à virus ébola en afrique de l'ouest: le point sur la situation et leçons apprises. In *Épidémie de maladie à Virus Ébola en Afrique de l'Ouest: le point sur la situation et leçons apprises*. O.M.S (2014)
2. Beeching, N.J., Fenech, M., Houlihan, C.F.: Ebola virus disease. BMJ **349**, (2014)
3. Gatherer, D.: The 2014 Ebola virus disease outbreak in West Africa. J. Gen. Virol. **95**(8), 1619–1624 (2014)
4. Fauci, A.S., Daszak, P., Morens, D.M.: The emergence of zoonotic diseases: the Ebola example. Emerg. Infect. Dis. **6**(4), 475–479 (2000)
5. C.D.C.: Épidémie de maladie à virus Ébola-afrique de l'ouest (2014). *CDC*. https://www.cdc.gov/mmwr/preview/mmwrhtml/mm6325a4_french.htm. Accessed 22 Jul 2024
6. Boulaasair, L., Bouzahir, H., Yavuz, M.: Global mathematical analysis of a patchy epidemic model. Int. J. Optim. Control Theor. Appl. (IJOCTA) **14**(4), 365–377 (2024)
7. Fandio, R., Abboubakar, H., Banbeto, S.A.G., Ekobena Fouda, H.P.: Mathematical modeling of the Cholera epidemic in Cameroon: from classical-order derivative to fractional-order derivative. MJ-MCS **1**(1), 1–32 (2025)
8. Abboubakar, H., Kumar, P., Erturk, V.S., Kumar, A.: A mathematical study of a tuberculosis model with fractional derivatives. Int. J. Model. Simul. Sci. Comput. **12**(4), 2150037 (2021)
9. Tsanou, B., Lubuma, J.M.-S., Moremedi, G.M., Morris, N., Kondera-Shava, R.: A simple mathematical model for Ebola in Africa. J. Biol. Dyn. **11**(1), 42–74 (2017)
10. Abboubakar, H., Racke, R.: Mathematical modeling of the coronavirus (covid-19) transmission dynamics using classical and fractional derivatives. Discrete Contin. Dyn. Syst. B **30**(1), 289–329 (2025)
11. Takgil, B., Kara, R.: Proposing a novel mathematical model for hospital pneumatic system. Int. J. Optim. Control Theor. Appl. (IJOCTA) **14**(2), 113–122 (2024)
12. Gündoğdu, H., Joshi, H.: Numerical analysis of time-fractional cancer models with different types of net killing rate. Mathematics **13**(3), 536 (2025)
13. Joshi, H.: Mechanistic insights of covid-19 dynamics by considering the influence of neurodegeneration and memory trace. Phys. Scr. **99**(3), 035254 (2024)
14. Joshi, H., Jha, B.K.: Fractional reaction diffusion model for parkinson's disease. In: International Conference on ISMAC in Computational Vision and Bio-Engineering, pp. 1739–1748. Springer (2018)
15. Yavuz, M., ur Rahman, M., Yildiz, M., Joshi, H.: Mathematical modeling of middle east respiratory syndrome coronavirus with bifurcation analysis. Contemp. Math. 3997–4012 (2024)
16. Naik, P.A., Yeolekar, B.M., Qureshi, S., Yeolekar, M., Madzvamuse, A.: Modeling and analysis of the fractional-order epidemic model to investigate mutual influence in HIV/HCV co-infection. Nonlinear Dyn. **112**(13), 11679–11710 (2024)
17. Din, R.U., Shah, K., Alqudah, M.A., Abdeljawad, T., Jarad, F.: Mathematical study of SIR epidemic model under convex incidence rate. AIMS Math. **5**(6), (2024)
18. Adel-Aissanou, K., Lekadir, O., Aïssani, D.: Modélisation mathématique de l'évolution de la pandémie covid-19: synthèse de la littérature. Séminaire Mathématique de Béjaia **18**(1), 9–27 (2020)
19. Abboubakar, H., Banbeto, S.A.G., Jan, R., Boulaaras, S.: Qualitative and quantitative analysis of the transmission dynamics of Ebola with convex incidence rates: a case study of Guinea. MESE 1–12 (2024)

20. Adu, I.K., Wireko, F.A., Nana-Kyere, S., Appiagyei, E., Osman, M.A., Asamoah, J.K.K.: Modelling the dynamics of Ebola disease transmission with optimal control analysis. MESE 1–27 (2024)
21. Agusto, F.B.: Mathematical model of Ebola transmission dynamics with relapse and reinfection. Math. Biosci. **283**, 48–59 (2017)
22. EL Rhoubari, Z., Besbassi, H., Hattaf, K., Yousfi, N.: Mathematical modeling of Ebola virus disease in bat population. Discrete Dyn. Nat. Soc. **2018**(1), 5104524 (2018)
23. Feng, Z., Zheng, Y., Hernandez-Ceron, N., Zhao, H., Glasser, J.W., Hill, A.N.: Mathematical models of Ebola–consequences of underlying assumptions. Math. Biosci. **277**, 89–107 (2016)
24. Jiang, S., Wang, K., Li, C., Hong, G., Zhang, X., Shan, M., Li, H., Wang, J.: Mathematical models for devising the optimal Ebola virus disease eradication. J. Transl. Med. **15**, 1–10 (2017)
25. Mbah, G.C.E., Onah, I.S., Ahman, Q.O., Collins, O.C., Asogwa, C.C., Okoye, C.: Mathematical modelling approach of the study of Ebola virus disease transmission dynamics in a developing country. Afr. J. Infect. Dis. **17**(1), 10–26 (2023)
26. Mishra, A., Mishra, B.K., Keshri, A.K.: Quarantine model on the transmission of Ebola virus disease in the human population with infectious dead class. Int. J. Appl. Comput. Math. **9**(6), 133 (2023)
27. Nazir, A., Ahmed, N., Khan, U., Mohyud-Din, S.T., Nisar, K.S., Khan, I.: An advanced version of a conformable mathematical model of Ebola virus disease in Africa. Alex. Eng. J. **59**(5), 3261–3268 (2020)
28. Ren, H., Xu, R.: Prevention and control of Ebola virus transmission: mathematical modelling and data fitting. J. Math. Biol. **89**(2), 25 (2024)
29. Mamo, D.K., Koya, P.R.: Mathematical modeling and simulation study of seir disease and data fitting of Ebola epidemic spreading in west Africa. J. Multidiscip. Eng. Sci. Technol. **2**(1), 106–114 (2015)
30. Tulu, T.W., Tian, B., Wu, Z.: Mathematical modeling, analysis and Markov Chain Monte Carlo simulation of Ebola epidemics. Results Phys. **7**, 962–968 (2017)
31. Bichara, D.: Étude de modèles épidémiologiques: Stabilité, observation et estimation de paramètres, Université de Lorraine (2013). PhD thesis
32. Aydın, S., Altun, K.: Early prediction of fabric quality using machine learning to reduce rework in manufacturing processes. Int. J. Optim. Control Theor. Appl. (IJOCTA) **14**(4), 308–321 (2024)
33. Selvamuthu, D., Khichar, D., Kalita, P., Jain, V.: Estimation of mortality rate of COVID-19 in India using SEIRD model. Opsearch 1–15 (2021)
34. Ahmad, A., Alzaidi, K., Sari, M., Uslu, H.: Prediction of anemia with a particle swarm optimization-based approach. Int. J. Optim. Control Theor. Appl. (IJOCTA) **13**(2), (2023)
35. Quintero, Y., Ardila, D., Camargo, E., Rivas, F., Aguilar, J.: Machine learning models for the prediction of the SEIRD variables for the covid-19 pandemic based on a deep dependence analysis of variables. Comput. Biol. Med. **134**, 104500 (2021)
36. Upadhyay, D., Joshi, H.: Mathematical modeling of local calcium signaling in neurons using artificial neural networks. Discrete Contin. Dyn. Syst. S **18**(5), 1392–1415 (2025)
37. Van den Driessche, P., Watmough, J.: Reproduction numbers and sub-threshold endemic equilibria for compartmental models of disease transmission. Math. Biosci. **180**(1–2), 29–48 (2002)
38. Kumari, A., Sood, M., et al.: Implementation of SimpleRNN and LSTMs based prediction model for coronavirus disease (covid-19). In: IOP Conference Series. Materials Science and Engineering, vol. 1022, p. 012015. IOP Publishing (2021)
39. Bahri, S., Kdayem, M., Zoghlami, N.: Deep learning for covid-19 prediction. In: 2020 4th International Conference on Advanced Systems and Emergent Technologies (IC_ASET), pp. 406–411. IEEE (2020)
40. Hassan, A., Shahin, I., Alsabek, M.B.: Covid-19 detection system using recurrent neural networks. In: CCCI 2020, pp. 1–5. IEEE (2020)
41. Feng, L.: SEIR model combined with LSTM and GRU for the trend analysis of COVID-19. Middle Tennessee State University (2021)

42. Ouedraogo, I.: Automatisation du pré-traitement des données par l'optimisation métaheuristique, Université du Québec à Chicoutimi (2023). PhD thesis
43. Sharma, M.: What steps should one take while doing data preprocessing. Hackernoon (2018)
44. Butcher, J.C.: NM-ODEs. John Wiley & Sons (2016)
45. Hodson, T.O.: Erreur quadratique moyenne (rmse) ou erreur absolue moyenne (mae): quand les utiliser ou non. Discussions sur le développement de modèles géoscientifiques **1–10**, 2022 (2022)
46. Chicco, D., Warrens, M.J., Jurman, G.: Le coefficient de détermination R-carré est plus informatif que SMAPE, MAE, MAPE, MSE et RMSE dans l'évaluation de l'analyse de régression. PeerJ Comput. Sci. **7**, e623 (2021)

Frontiers in HIV-1 Dynamic Modeling: A Neural and Delay-Driven Perspective

P. Krishnapriya

Abstract Accurate modeling of HIV-1 infection dynamics necessitates the incorporation of biologically meaningful time delays that represent critical stages of viral replication and immune response. This chapter introduces a novel computational framework that integrates time-delayed neural networks (TDNNs) with the Levenberg-Marquardt backpropagation (LMB) optimization algorithm, referred to as TDNNs-LMB, for solving a delay differential equation–based HIV-1 dynamic model. The proposed approach exploits the strength of TDNNs in capturing temporal dependencies and nonlinear system behavior, while utilizing the rapid convergence and numerical stability offered by the LMB algorithm. The dataset is partitioned into training (70%), validation (15%), and testing (15%) subsets to ensure robust learning and generalization. A reference numerical solution is employed for benchmarking, and the model performance is evaluated through absolute error (AE), mean square error (MSE), correlation coefficients, and a suite of graphical diagnostics, including fitting curves, state-transition diagrams, and error histograms. The results demonstrate that the TDNNs-LMB framework achieves high predictive accuracy, robustness, and computational efficiency, outperforming traditional numerical methods. The proposed methodology offers a powerful and flexible tool for investigating delay-induced effects in HIV-1 progression and is readily extensible to a broad class of biological and biomedical systems with time-dependent dynamics.

Keywords HIV-1 infection models · Delay differential equations · Time-delayed neural networks · Levenberg-marquardt backpropagation

---

P. Krishnapriya (✉)
Department of Mathematics, Thigarajar College of Engineering, Madurai, India
e-mail: priyaprithu1205@gmail.com

L. Kaur and P. Kumar (eds.), *Neural Networks: Theory, Algorithms, Simulation, and Applications*, Studies in Computational Intelligence 1273,
https://doi.org/10.1007/978-3-032-18750-5_10

1 Introduction

Over the past three decades, mathematical modeling has been instrumental in deepening scientific understanding of HIV-1 infection dynamics. Early studies based on systems of ordinary differential equations (ODEs) provided fundamental insights into viral replication kinetics, $CD4^+$ T-cell depletion, and the ability of antiretroviral therapy (ART) to suppress infection and prolong immune function. These foundational models were progressively enriched to account for diverse biological phenomena, including immune effector cell responses, latent reservoirs, mutational drug resistance, and structured treatment interruptions, thereby producing increasingly accurate descriptions of disease progression and therapeutic outcomes [1, 2].

A major advancement in the field came with the introduction of delay differential equation (DDE) models, which explicitly account for inherent time lags in the viral life cycle such as intracellular eclipse periods, viral production delays, and delayed immune activation cycles. Incorporating such delays significantly enhances biological realism and predictive fidelity, especially in long-term simulations of viral load oscillations, immune exhaustion, and treatment rebound [3–5]. These models have provided a more mechanistic understanding of how temporal processes shape infection persistence and control thresholds, continuing to influence both theoretical investigations and clinical decision-making.

Parallel to these developments, the rapid growth of computational intelligence and machine-learning methodologies has opened new avenues for complementing traditional mathematical approaches. Neural network–based techniques, particularly those designed to capture temporal structures and nonlinear relationships, have shown remarkable promise for learning hidden dynamics from sparse or noisy clinical data. When combined with mechanistic delay models, neural optimization frameworks can improve predictive accuracy, curve-fitting capability, and parameter identifiability, enabling deeper exploration of viral replication patterns and patient-specific disease trajectories [6, 7].

A central component of neural optimization research has been the Levenberg–Marquardt (LM) algorithm and its derivatives, owing to their robustness and efficiency in training medium-scale networks. However, the classical LM method often converges to local minima, motivating numerous researchers to enhance or hybridize the technique. Many recent studies have focused on guiding the LM algorithm toward more favorable local solutions without the need to identify an actual global minimum which is typically unnecessary in modern neural network applications. To address this, hybridization strategies have been employed: genetic algorithms have been used to generate improved initial weights for LM networks [8, 9], Bayesian estimation has been incorporated to calibrate model parameters, and wavelet-based preprocessing has been combined with LM to improve data representation and learning quality [10, 11]. Other studies have integrated least-squares support vector machines (LSSVM) with LM, forming hybrid learning architectures that achieve better generalization and computational performance [12]. These hybrid frameworks have consistently demonstrated strong empirical effectiveness across engineering and biomedical domains.

Beyond hybrid techniques, several researchers have proposed enhancements that modify the LM algorithm itself. Yang [13] introduced a high-order LM scheme exhibiting biquadratic convergence, significantly accelerating convergence rates. Chen [14] designed an accelerated fourth-order LM variant that further improves computational efficiency. Derakhshandeh [15] developed a novel three-step LM algorithm grounded in fuzzy logic theory, offering improved stability and convergence behavior. Another notable contribution is the adaptive LM-based echo state network proposed in [16], which incorporates a dynamically adjusted damping factor to better manage error landscapes. These variations, validated through trust-region techniques, collectively demonstrate the growing sophistication of LM-based optimization methods and their potential for high-accuracy model training and system identification. Furthermore, Kumar [17] extended the modelling framework by considering integer- and fractional-order nonlinear HIV-1 models supported by real-data fitting. Their work utilized the Caputo fractional derivative for fractional-order analysis and applied neural network techniques to examine the complex dynamical behaviour of the system.

Within this broader context, the present chapter introduces a biologically informed DDE model of HIV-1 infection incorporating key intracellular and immune-mediated delays. To address the analytical and numerical challenges posed by delay systems, we develop a deep learning framework based on a time-delay neural network trained using an optimized LM-type methodology. To the best of our knowledge, this work represents the first application of a time-delay neural architecture, optimized using a modern LM variant, to HIV-1 DDE models. The proposed hybrid modeling strategy achieves excellent agreement with analytical benchmarks, demonstrates high prediction robustness, and offers a systematic error and convergence analysis.

Beyond methodological novelty, this integrated approach provides an advanced computational platform for studying viral dynamics, estimating key biological delays, exploring immune control thresholds, forecasting treatment responses, and evaluating alternative therapeutic strategies. By bringing together delay-based mathematical models and neural optimization frontiers particularly the evolving LM-based family of algorithms this chapter contributes a unified framework that advances both theoretical modeling and practical clinical understanding of HIV-1 infection.

2 HIV-I Mathematical Model

During the course of viral infection, the host immune defense transitions from an immediate, nonspecific innate response to a delayed, antigen-specific cellular response. The present study investigates the intricate population dynamics among viral particles, host $CD4^+$ T cells, and cytotoxic T lymphocyte (CTL) responses. To account for the intrinsic time delays associated with viral replication, immune activation, and therapeutic intervention, the system is formulated as a set of delay differential equations (DDEs), allowing for a comprehensive characterization of infection dynamics and immune regulation.

The model describes the interactions among uninfected target cells T, infected cells T^*, free virus particles V, and immune effector cells E. Activated $CD4^+$ T cells are produced at a constant rate λ (cells/day) and die at a rate d (day^{-1}). These cells become infected by free virions at a rate k (day^{-1}cell^{-1}), yielding infected cells T^* that are cleared at a rate δ (day^{-1}). The CTL response contributes to the elimination of infected cells with a killing rate constant d_X (day^{-1}cell^{-1}). Free virus particles are generated by infected cells at a rate N and decay at a rate c (day^{-1}). The CTL population expands in response to infection at a rate p and decays naturally at a rate d_E (day^{-1}). The resulting model system is formulated as follows:

$$
\begin{aligned}
\dot{T} &= \lambda - dT - kTV, \\
\dot{T^*} &= kT(t-\tau_1)V(t-\tau_1) - \delta T^* - d_X E T^*, \\
\dot{V} &= N\delta T^*(t-\tau_2) - cV, \\
\dot{E} &= pT^*(t-\tau_3) - d_E E.
\end{aligned}
\tag{1}
$$

It incorporates discrete time delays, where τ_1 denotes the intracellular delay during infection, τ_2 represents the time required for virus production, and τ_3 corresponds to the delay in immune response activation. This model provides a rigorous framework for analyzing the effects of intrinsic time delays on viral infection dynamics and for evaluating the efficacy of antiviral treatment strategies.

The dynamics of HIV-1 infection are inherently nonlinear and time-delayed, as viral replication, immune response activation, and infected $CD4^+$ T cell dynamics occur over distinct time scales. Modeling such systems using DDEs introduces complex temporal dependencies, making parameter estimation and accurate prediction challenging. LMB algorithm is particularly suited for this problem because it efficiently minimizes the error between model predictions and observed data, even in the presence of nonlinearities and time delays. By iteratively adjusting network weights or model parameters, the LMB method ensures that the TDNN-based DDE model can accurately capture delayed viral proliferation, immune responses, and other nonlinear behaviors. This precision is critical for reliable simulation of disease progression and for evaluating the potential impact of therapeutic interventions.

2.1 Prerequisite

We denote by $\mathbb{R}^4$ the four-dimensional real Euclidean space equipped with the norm $|.|$. For $\tau > 0$, let $C = C([-\tau, 0], \mathbb{R}^4_+)$ be the Banach space of continuous functions mapping $[-\tau, 0]$ into $\mathbb{R}^4_+$, endowed with the topology of uniform convergence, where $\tau = \max\{\tau_1, \tau_2, \tau_3\}$.

According to the classical theory of functional differential equations [18–20], for any initial function $\phi \in C([-\tau, 0], \mathbb{R}^4_+)$, there exists a unique solution

$$Z(t, \phi) = (T(t, \phi), T^*(t, \phi), V(t, \phi), E(t, \phi))$$

of the delayed system (1) satisfying $Z_0 = \phi$. The initial function $\phi = (\phi_1, \phi_2, \phi_3, \phi_4)$ is assumed to satisfy $\phi_i(\xi) \geq 0$ for $\xi \in [-\tau, 0]$ and $\phi_i(0) > 0$ for $i = 1, \ldots, 4$. Consequently, the initial conditions can be written as

$$T(\xi) = \phi_1(\xi), \quad T^*(\xi) = \phi_2(\xi), \quad V(\xi) = \phi_3(\xi), \quad E(\xi) = \phi_4(\xi), \quad \xi \in [-\tau, 0]. \quad (2)$$

2.2 *Novel Features of the TDNN–LMB Framework for the DDE-HIV-1 Model*

1. The TDNN architecture inherently captures temporal dependencies, enabling accurate modeling of biological delays in viral replication, infected $CD4^+$ T cell dynamics, and immune responses.
2. LMB algorithm provides robust and efficient parameter optimization, ensuring reliable convergence despite the strong nonlinearities and time delays present in HIV-1 dynamics.
3. The framework effectively handles complex nonlinear interactions among virus particles, host cells, and immune effectors while explicitly incorporating discrete time delays.
4. Integration of TDNNs with LMB optimization yields high prediction accuracy for V and T^* enhancing the reliability of disease progression simulations.
5. The approach allows simulation of various therapeutic interventions, facilitating evaluation of treatment strategies and their effects on HIV-1 dynamics.
6. Although demonstrated for HIV-1, the TDNN–LMB framework is generalizable to other infectious disease models and complex dynamic systems involving time delays.

3 Methodology: LMB-Optimized Time-Delay Neural Network for HIV-1 DDE Modeling

This section presents the computational framework designed to solve the delay-driven HIV-1 infection model using a Time-Delay Neural Network (TDNN) optimized with the Levenberg–Marquardt Backpropagation (LMB) algorithm. The methodology integrates neural approximation, delay embedding, residual minimization, and trust-region optimization into a unified pipeline that yields accurate and robust solutions for nonlinear biological systems with inherent temporal delays.

3.1 Neural Approximation of the HIV-1 Delay Differential Model

We consider the HIV-1 system with biologically relevant delays:

$$\begin{aligned}
\dot{T}(t) &= \lambda - dT(t) - kT(t)V(t),\\
\dot{T^*}(t) &= kT(t-\tau_1)V(t-\tau_1) - \delta T^*(t) - d_X E(t)T^*(t),\\
\dot{V}(t) &= N\delta T^*(t-\tau_2) - cV(t),\\
\dot{E}(t) &= pT^*(t-\tau_3) - d_E E(t).
\end{aligned} \tag{3}$$

Each state variable is approximated by a neural structure of the form

$$\widehat{X}(t; \mathbf{w}) = \sum_{j=1}^{m} w_j \sigma(a_j t + b_j) + b_0, \tag{4}$$

where $\sigma(\cdot)$ is a smooth activation function, m is the number of hidden neurons, and $\mathbf{w}$ denotes the weight vector.

Delay terms such as $T(t-\tau_1)$ and $V(t-\tau_1)$ are implemented using explicit time-shift connections via TDNN delay taps. Neural derivatives $\dot{\widehat{X}}(t)$ are evaluated using automatic differentiation.

3.2 Construction of Residual Dynamics

To ensure that the neural approximations satisfy the DDE structure, residual functions are constructed:

$$R_T(t) = \dot{\widehat{T}}(t) - \left(\lambda - d\widehat{T}(t) - k\widehat{T}(t)\widehat{V}(t)\right), \tag{5}$$

$$R_{T^*}(t) = \dot{\widehat{T^*}}(t) - \left(k\widehat{T}(t-\tau_1)\widehat{V}(t-\tau_1) - \delta\widehat{T^*}(t) - d_X\widehat{E}(t)\widehat{T^*}(t)\right), \tag{6}$$

$$R_V(t) = \dot{\widehat{V}}(t) - \left(N\delta\widehat{T^*}(t-\tau_2) - c\widehat{V}(t)\right), \tag{7}$$

$$R_E(t) = \dot{\widehat{E}}(t) - \left(p\widehat{T^*}(t-\tau_3) - d_E\widehat{E}(t)\right). \tag{8}$$

The total loss functional is expressed as

$$\mathcal{L}(\mathbf{w}) = \sum_{t\in\Omega}\left[R_T(t)^2 + R_{T^*}(t)^2 + R_V(t)^2 + R_E(t)^2\right] + \beta\mathcal{L}_{\mathrm{IC}}, \tag{9}$$

where β is a penalty parameter enforcing consistency with initial history functions.

3.3 *Levenberg–Marquardt Backpropagation for Parameter Optimization*

The LMB algorithm minimizes the loss functional (9) by iteratively updating the weight vector $\mathbf{w}$ according to the nonlinear least-squares update rule:

$$\mathbf{w}_{k+1} = \mathbf{w}_k - \left(J_k^T J_k + \mu_k I\right)^{-1} J_k^T \mathbf{R}_k, \tag{10}$$

where J_k is the Jacobian matrix of residuals, $\mathbf{R}_k$ is the residual vector, and μ_k is the damping parameter.

Damping Strategy (Trust-Region Logic)

- $\mathcal{L}(\mathbf{w}_{k+1}) < \mathcal{L}(\mathbf{w}_k)$: reduce μ_k (algorithm behaves like Gauss–Newton).
- $\mathcal{L}(\mathbf{w}_{k+1}) \geq \mathcal{L}(\mathbf{w}_k)$: increase μ_k (algorithm becomes gradient-descent-like).

This adaptive switching yields fast convergence and stability even in the presence of delay-induced oscillations and nonlinear immune–virus interactions.

4 Numerical Simulations

To assess the predictive accuracy and robustness of the TDNN–LMB framework for the DDE-based HIV-1 model, the network was trained using time-series data of T, T^*,V,E. Its performance was evaluated through multiple metrics, including regression analysis, error distributions, and mean squared error trajectories, to verify the model's ability to capture nonlinear dynamics and delayed immune responses. The parameters values are taken from [5].

The neural network was trained on time-series data of infected and uninfected populations, viral load, and recovered individuals to capture the nonlinear dynamics of HIV-1. Predicted outputs closely match actual data across training, validation, and testing Fig. 1a–g, with $R = 1$, low errors, smooth convergence, and accurate curve fitting for $T(t)$, $I(t)$, $V(t)$, and $R(t)$. Absolute errors remain within acceptable limits, confirming the model's reliability, stability, and suitability for biological forecasting and intervention analysis.

The predictive performance and learning behavior of the TDNN–LMB model for the DDE-based HIV-1 system were further analyzed through key training metrics and graphical evaluations. These visual diagnostics collectively demonstrate the model's learning efficiency, convergence stability, and predictive precision.

The network exhibited smooth convergence with minimal oscillations and stable learning, as illustrated in Fig. 2a–g, indicating robustness and resistance to overfitting. The mean squared error declined rapidly to a low optimal value, while regression

Algorithm 1 Time-Delayed Neural Network Trained with Levenberg–Marquardt Algorithm for Eq. (1)

Require: Sampled data $\mathbf{x}(t_k) = [T, T^*, V, E]^T$, delays τ_1, τ_2, τ_3, sampling step Δt, and LM parameters $(\mu_0, \mu_{\text{inc}}, \mu_{\text{dec}}, \alpha, \text{tol}, \text{max\_iter})$
Ensure: Trained weights w^* approximating $\dot{\mathbf{x}}(t)$
Data Preparation:
Simulate Eq. (1) or collect $\mathbf{x}(t_k)$ on $[t_0, t_f]$.
Compute $\mathbf{y}_k = \dot{\mathbf{x}}(t_k)$ numerically.
Construct delayed inputs:

$$\mathbf{u}_k = [T(t_k), T^*(t_k), V(t_k), E(t_k), T(t_k - \tau_1), V(t_k - \tau_1), T^*(t_k - \tau_2), T^*(t_k - \tau_3)]^T.$$

Initialization:
Initialize network weights w randomly.
Set $\mu \leftarrow \mu_0$, iter $\leftarrow 0$.
while iter < max_iter **do**
  Compute predicted outputs $\hat{\mathbf{y}}_k = \text{NN}(\mathbf{u}_k; w)$.
  Compute residuals $\mathbf{r}_k = \hat{\mathbf{y}}_k - \mathbf{y}_k$.
  Compute cost $L = \frac{1}{2}\sum_k \|\mathbf{r}_k\|^2 + \frac{\alpha}{2}\|w\|^2$.
  Compute Jacobian $J = \partial \mathbf{r}/\partial w$ by backpropagation.
  Solve $(J^T J + \mu I)\Delta w = -J^T \mathbf{r}$.
  $w_{\text{cand}} \leftarrow w + \Delta w$.
  Evaluate new cost L_{cand}.
  if $L_{\text{cand}} < L$ **then**
    $w \leftarrow w_{\text{cand}}$; $\mu \leftarrow \mu/\mu_{\text{dec}}$.
  else
    $\mu \leftarrow \mu \cdot \mu_{\text{inc}}$.
  end if
  if $\|\Delta w\| < \text{tol}(\|w\| + \text{tol})$ **then**
    break
  end if
  iter $\leftarrow$ iter + 1
end while
Validation:
Integrate the DDE with $f_{\text{NN}}(\mathbf{u}; w^*)$ and compare trajectories.
return w^*

analysis yielded $R = 1$ across all datasets. Error histograms were tightly centered around zero, confirming low bias and high prediction accuracy. Predicted trajectories for $T(t)$, $I(t)$, $V(t)$, and $R(t)$ closely matched experimental data within narrow 95% confidence intervals, capturing target cell depletion, viral rise and fall, and recovery trends. These results demonstrate the model's reliability and suitability for forecasting HIV-1 progression and supporting intervention strategies.

To further evaluate the predictive capability and biological consistency of the proposed TDNN–LMB model, temporal and statistical analyses were conducted on the simulated outputs. These assessments provide a comprehensive understanding of the model's ability to reproduce key infection dynamics and its robustness against biological variability (Table 1).

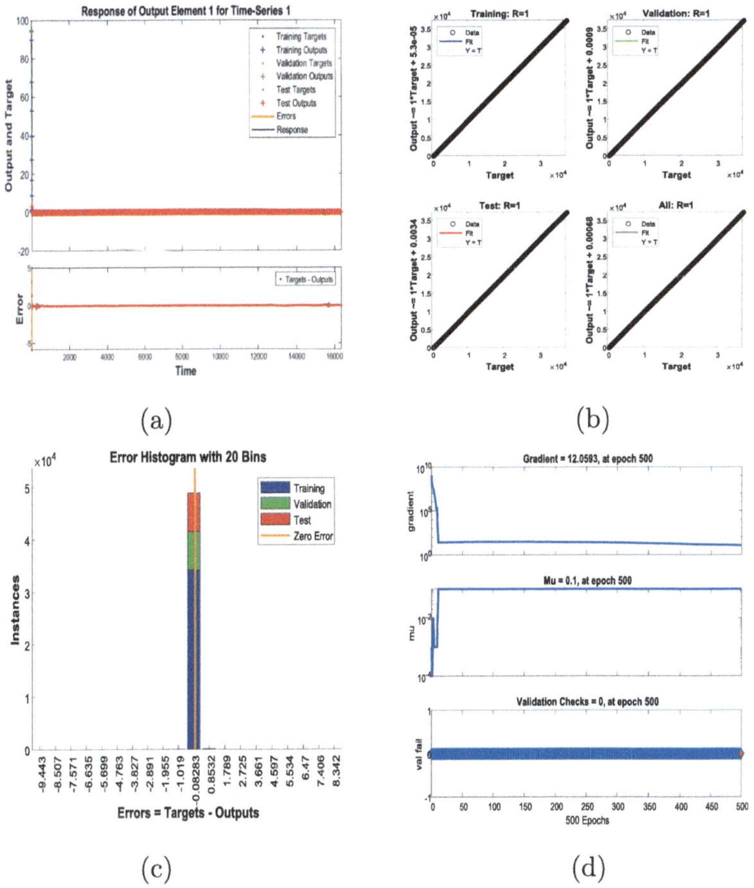

Fig. 1 Performance evaluation of the proposed TDNN model for Case 1, showing response, regression, error analysis, training performance, validation and curve fitting results

Figure 3a–g shows the temporal prediction curves for $T(t)$, $I(t)$, $V(t)$, and $R(t)$ show strong agreement with experimental data, with narrow 95% confidence bounds and consistently low absolute errors across training, validation, and testing. The model accurately captures infection dynamics, including target cell depletion, viral rise–fall behavior, and recovery trends, reflecting biological realism. Statistically, a perfect correlation ($R = 1$) and zero-centered, symmetric error histograms confirm unbiased and precise predictions. The training process exhibited smooth convergence and stable learning, with MSE rapidly reaching a minimal value, indicating strong generalization and resistance to overfitting. Overall, these results validate the model's reliability, stability, and suitability for forecasting HIV-1 progression and evaluating therapeutic strategies.

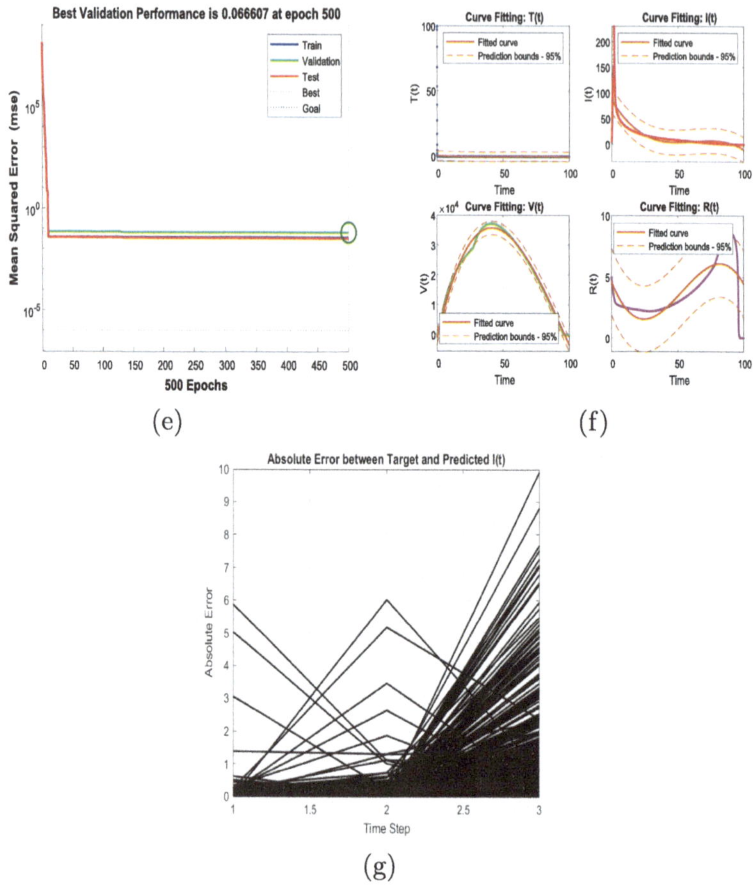

Fig. 1 (continued)

To assess the predictive capability of the TDNN-LMB model applied to the DDE–HIV-I framework, a comprehensive performance evaluation was conducted. Multiple statistical and graphical diagnostics were employed to validate the model's accuracy, stability, and biological consistency across all datasets.

Figure 4a–g exposes the performance of the proposed computational model was evaluated using diagnostic plots to assess accuracy, stability, and generalizability. Time-series responses showed close agreement between predicted and experimental data across training, validation, and testing, with minimal bounded errors and consistent long-term accuracy. Regression analysis yielded near-perfect correlations ($R_{train} = 0.99999$, $R_{val} = 0.99993$, $R_{test} = 0.99944$, $R_{all} = 0.99993$), indicating unbiased predictions and strong data fitting. The MSE decreased rapidly to its minimum (975.6092 at epoch 9), supported by stable gradients and low validation checks, confirming smooth convergence. Curve fitting results for $T(t)$, $I(t)$, $V(t)$,

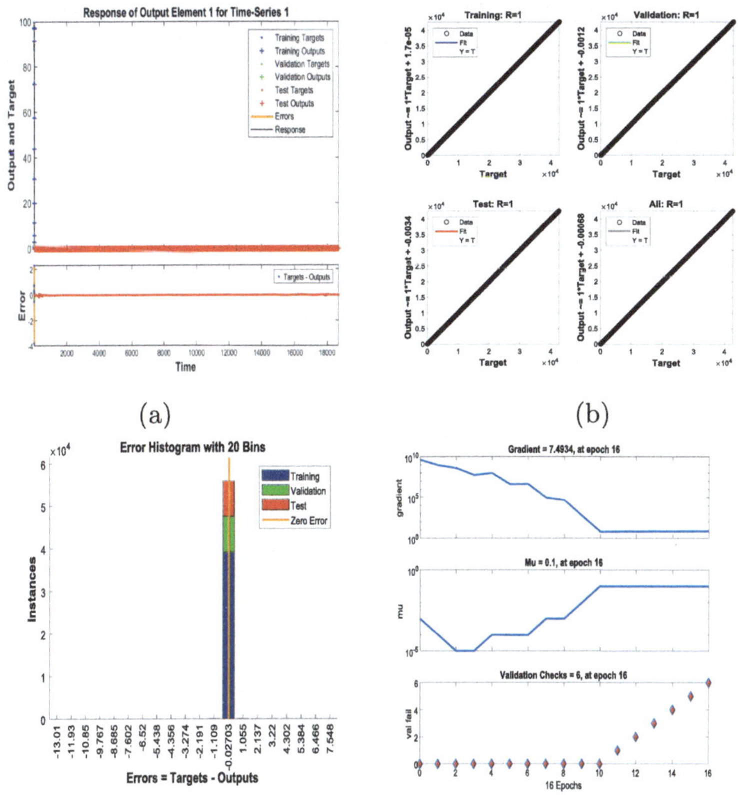

(a) (b)

(c) (d)

Fig. 2 Performance evaluation of the proposed TDNN model for Case 2, showing response, regression, error analysis, training performance, validation and curve fitting results

and $R(t)$ closely matched fitted curves within 95% bounds, accurately capturing both transient and steady-state dynamics. Absolute error analysis showed small deviations within acceptable limits, demonstrating the model's robustness, predictive reliability, and suitability for forecasting and control of complex biological systems.

To validate the effectiveness of the proposed TDNN–LMB model, we conducted comprehensive analyses of its predictive accuracy, training behavior, and robustness. Both temporal and statistical evaluations were performed to assess the model's ability to reproduce key biological dynamics and generalize across unseen data.

Figure 5a–g exposes graphical and temporal analyses demonstrate the strong predictive performance of the proposed model in capturing the dynamics of the biological system. Time-series predictions for target cells, viral load, recovered populations, and infected cells closely matched experimental data across training, validation, and test sets, with narrow 95 % confidence intervals and consistently low absolute errors. Scatter plot analyses yielded $R = 1$ in all phases, and error histograms were sym-

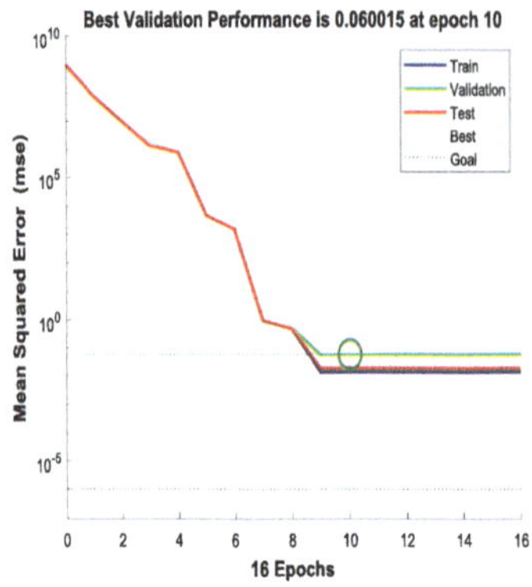

(e)

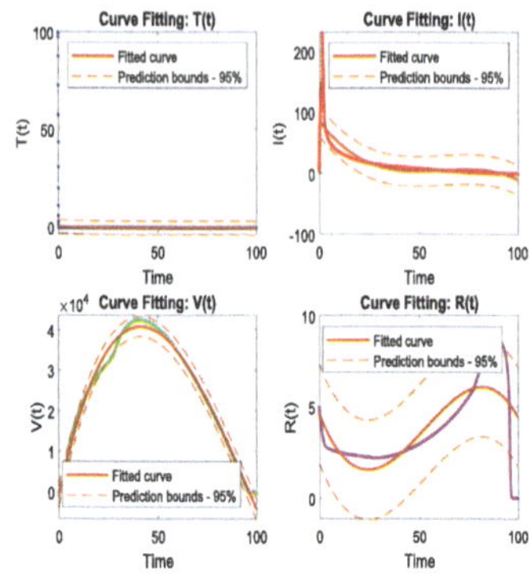

(f)

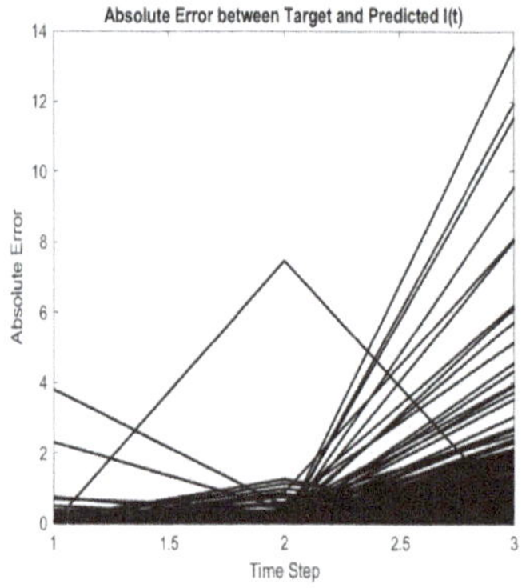

(g)

Fig. 2 (continued)

Table 1 Comparative summary of training progress across five neural network experiments

Unit	Case 1	Case 2	Case 3	Case 4	Case 5
Epoch (Stopped)	26	16	19	13	67
Elapsed time	00:00:39	00:03:40	00:00:56	00:00:01	00:00:06
Performance (MSE)	2.99×10^{-2}	1.67×10^{-2}	2.6×10^{-2}	5.49×10^{1}	1.79×10^{-1}
Gradient	88.1	247	137	8.42×10^{4}	39.8
Mu	0.01	0.01	0.01	1	0.1
Validation checks	6	6	6	6	6

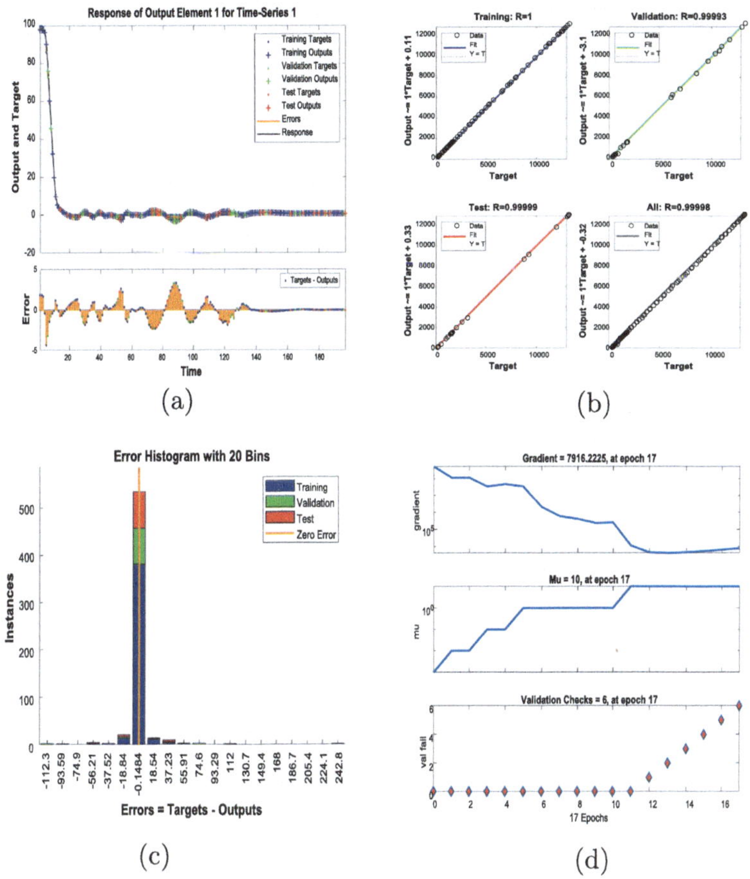

Fig. 3 Performance evaluation of the proposed TDNN model for Case 3, showing response, regression, error analysis, training performance, validation and curve fitting results

metric and zero-centered, confirming unbiased and accurate predictions. The training process showed smooth convergence, stable learning rates, minimal validation checks, and no overfitting, with MSE rapidly reaching its optimal value and maintaining test performance. Overall, the model reliably reproduces complex biological dynamics, generalizes to unseen data, and provides a robust framework for infectious disease modeling and therapeutic strategy assessment.

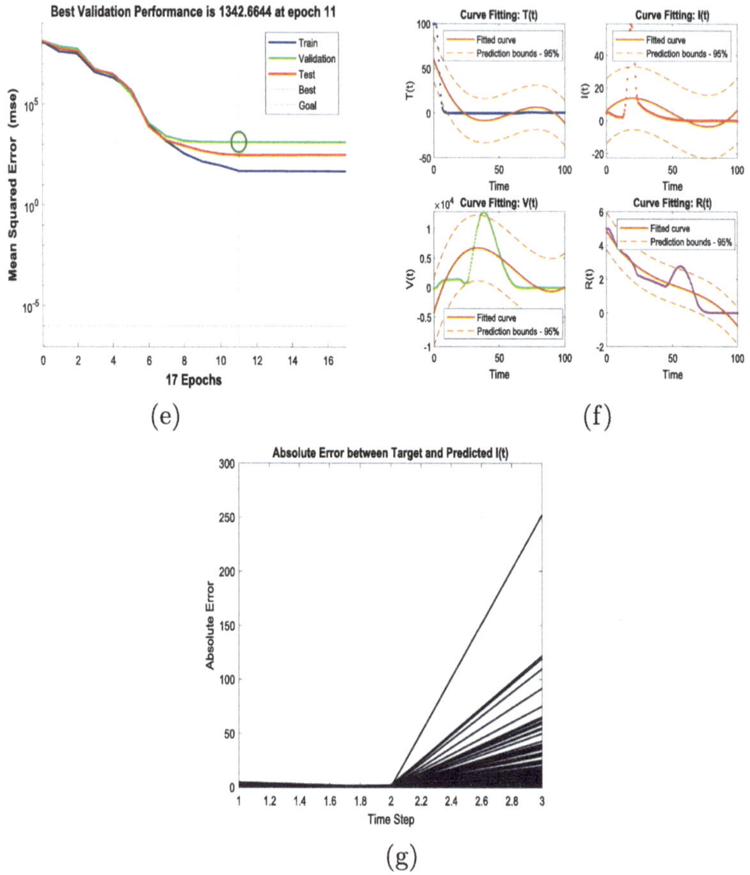

Fig. 3 (continued)

5 Numerical Implementation and Results Using TDNN–LMB for the HIV-1 DDE Model

This chapter presents the numerical findings obtained using Time-Delay Neural Networks (TDNNs) optimized with the Levenberg–Marquardt Backpropagation (LMB) algorithm for solving a nonlinear delay differential equation (DDE) model of Human Immunodeficiency Virus type-1 (HIV-1). The TDNN–LMB methodology was employed on numerically generated data derived from the HIV-1 DDE system, which describes the dynamic interactions among healthy CD4$^+$ T cells (T), infected cells (T^*), free virus particles (V), and immune effector cells (E). The model incorporates discrete intracellular delays as well as immune response delays to reflect biologically realistic infection pathways.

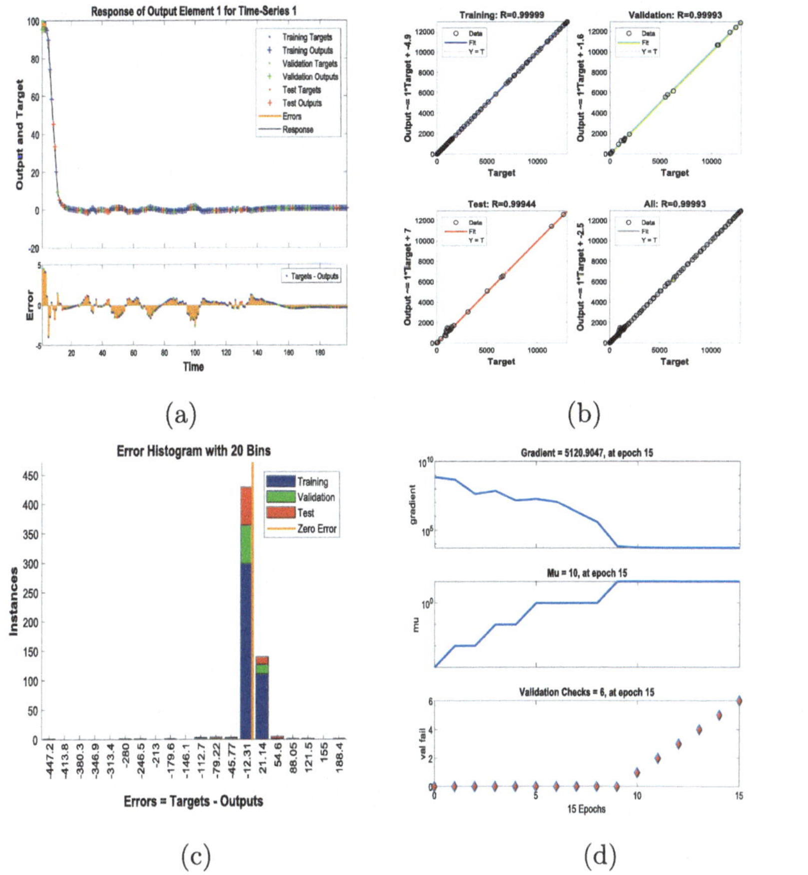

Fig. 4 Performance evaluation of the proposed TDNN model for Case 4, showing response, regression, error analysis, training performance, validation and curve fitting results

To ensure reliable learning and generalization, the constructed dataset was systematically divided into training, validation, and testing subsets. The TDNN–LMB framework was then evaluated across a range of biologically meaningful delay configurations. For each configuration, the reference trajectories obtained through numerical integration of the DDE system served as benchmark solutions against which the neural network approximations were assessed.

The results indicate a strong concordance between the TDNN predictions and the benchmark numerical solutions. This agreement is further supported by the observed convergence behaviour during training and the low absolute error profiles across all trajectories. Collectively, these findings demonstrate that the TDNN–LMB algorithm constitutes a robust computational tool for approximating and predicting the evolution of biologically delayed systems, particularly in the context of HIV-1 infection.

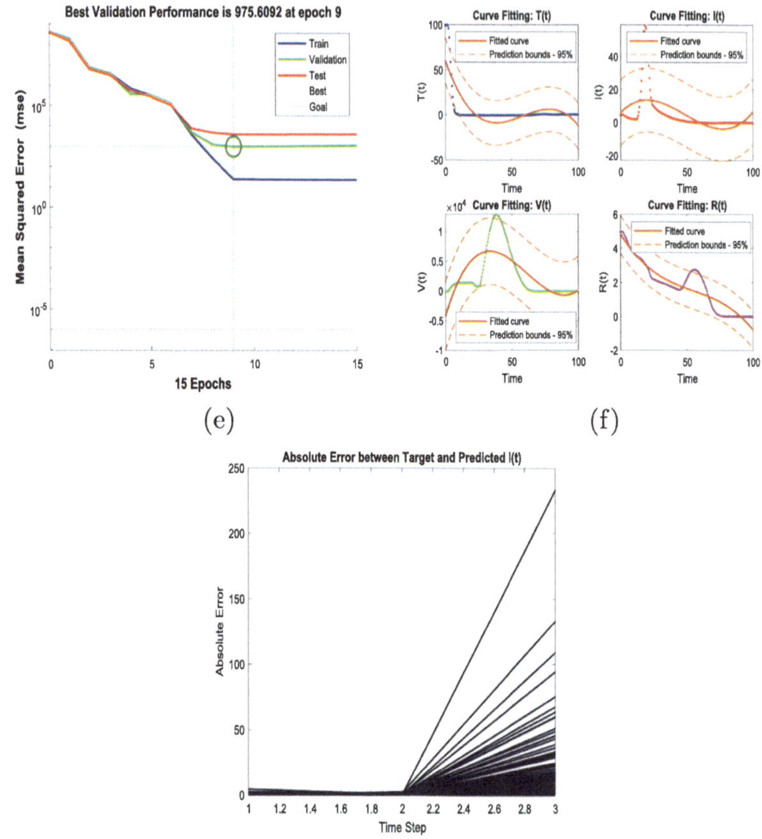

Fig. 4 (continued)

Beyond its accuracy, the approach offers considerable flexibility compared to classical numerical solvers. Its ability to rapidly generate high-quality predictions under varying physiological parameters and therapeutic scenarios makes it a valuable framework for exploring treatment strategies and understanding the progression of infection dynamics in systems with inherent biological delays.

6 Conclusion

This chapter has demonstrated that, a Time-Delay Neural Network trained using the Levenberg–Marquardt Backpropagation (TDNN–LMB) algorithm was successfully employed to model and predict the nonlinear delayed dynamics of the HIV-1

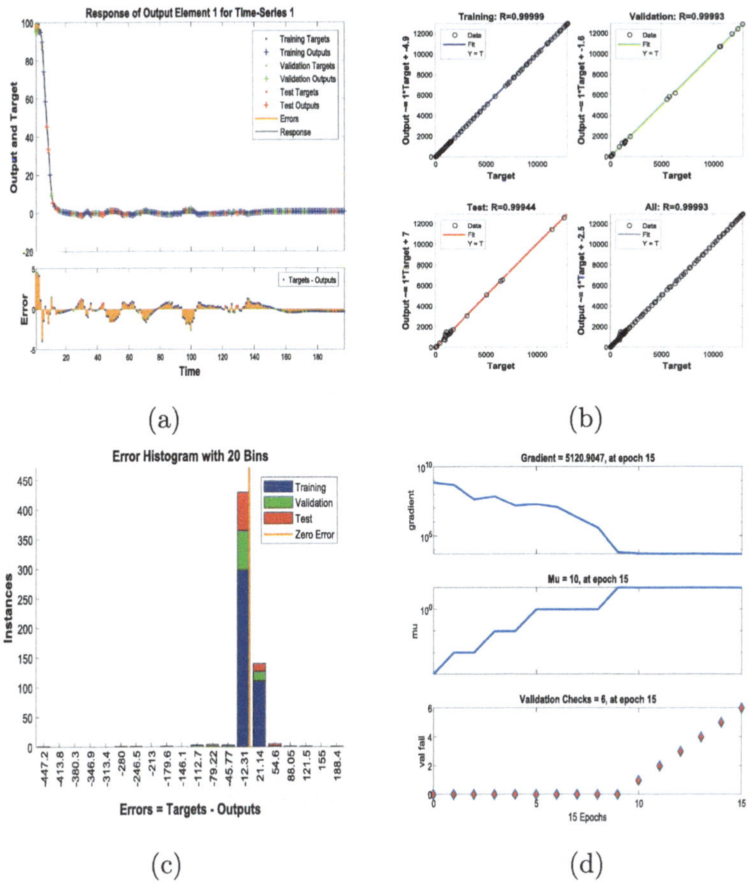

Fig. 5 Performance evaluation of the proposed TDNN model for Case 5, showing response, regression, error analysis, training performance, validation and curve fitting results

infection system. The proposed TDNN–LMB framework accurately reproduced the trajectories of T, T^*, V, and E, showing excellent consistency with the benchmark solutions generated by the delay differential equation (DDE) model. Quantitative performance measures, including the Mean Squared Error (MSE), Root Mean Squared Error (RMSE), and regression coefficients (R), further verified the reliability, efficiency, and precision of the neural network approach.

Overall, these findings demonstrate that the TDNN–LMB methodology provides a powerful and robust computational tool for simulating HIV-1 infection dynamics. Its ability to capture complex delayed interactions makes it a valuable complement to traditional numerical solvers. Beyond accurately approximating biological trajectories, the framework offers significant potential for accelerating exploratory studies, supporting therapeutic strategy design, and enhancing predictive modelling in systems governed by biological time delays. This chapter therefore establishes

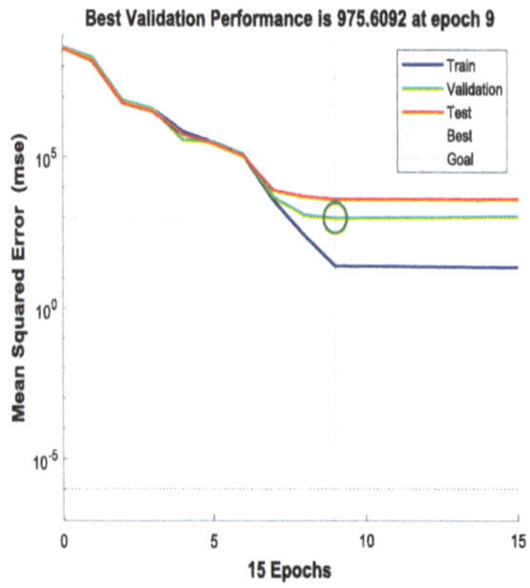

(e)

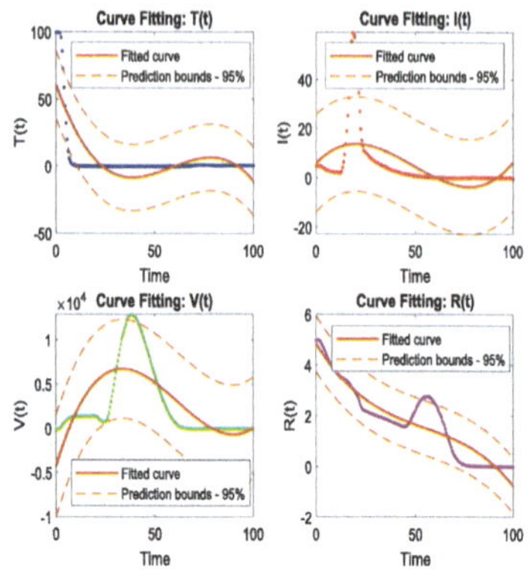

(f)

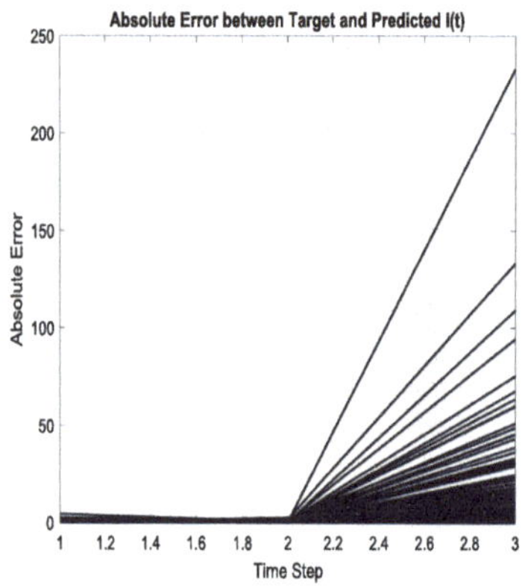

(g)

Fig. 5 (continued)

the TDNN–LMB approach as an effective and promising methodology for future research on biologically delayed dynamical systems.

Competing Interests The authors have no conflicts of interest to declare that are relevant to the content of this chapter.

References

1. Perelson, A.S., Nelson, P.W.: Mathematical analysis of HIV-1 dynamics in vivo. SIAM Rev. **41**, 3–44 (1999)
2. Perelson, A.S., Kirschner, D.E., De Boer, R.: Dynamics of HIV-1 infection of $CD4^{+}T$ cells. Math. Bio. Sci. **114**, 81–125 (1993)

3. Maheswari, M.C., Krishnapriya, P., Krishnan, K., Pitchaimani, M.: A mathematical model of HIV-1 infection within host cell to cell viral transmissions with RTI and discrete delays. J. Appl. Math. Comp. **56**, 151–178 (2018)
4. Monica, C., Pitchaimani, M.: Analysis of stability and Hopf bifurcation for HIV-1 dynamics with PI and three intracellular delays. Nonlinear. Anal. Real World. App. **27**, 55–69 (2016)
5. Pitchaimani, M., Monica, C.: Global stability analysis of HIV-1 infection model with three time delays. J. Appl. Math. Comp. **48**, 293–319 (2014)
6. Ali, D., Shah, F., Asadullah, Javeed, S., Ayesha, U., Abdullaeva, B.S., Gupta, M., Tarakaramu, N.: Numerical investigation of a fractional order differential human papillomavirus model using time delay neural network. Part. Diff. Equ. Appl. Math. **14**, 101196 (2025)
7. Bano, F., Suhail, H., Serbaya, A., Rizwan, M., Shabaz, F., Hasan, Hany, S., Khalifa: An artificial neural network and Levenberg-Marquardt training algorithm-based mathematical model for performance prediction. Appl. Math. Sci. Eng **32**, 151–178 (2024)
8. Zhao, L., Otoo, C.O.A.: Stability and complexity of a novel three-dimensional envi-ronmental quality dynamic evolution system. Complexity, 3941920 (2019)
9. Zhou, W., Liu, D., Hong, T.: Application of GA-LM-BP neural network in fault prediction of drying furnace equipment. Matec. Web Conf. **232**, 01041 (2018)
10. Jia, P., Zhang, P.: Type identification of coal mining face based onWavelet packet decomposition and LM-BP. In: Proceedings of the 2018 IEEE 9th International Conference on Software Engineering and Service Science (ICSESS), Beijing, China (2018)
11. Hua, L., Bo, L., Tong, L., Wang, M., Fu, H., Guo, R.: Angular acceleration sensor fault diagnosis based on LM-BP neural network. In: Proceedings of the 37th Chinese Control Conference, pp. 6028–6032. Wuhan, China (2018)
12. Hossein, A.M., Nazari, M.A., Madah, R.G.H., BehshadShafii, M., Ahmadi, M.A.: Thermal conductivity ratio prediction of Al_2O_3/water nanofluid by applying connectionist methods. Colloids Surf. A Physicochem. Eng. Asp. **541**, 154–164 (2018)
13. Yang, X.: A higher-order Levenberg-Marquardt method for nonlinear equations. Appl. Math. Comput. **219**, 10682–10694 (2013)
14. Chen, L.: A high-order modified Levenberg-Marquardt method for systems of nonlinear equations with fourth-order convergence. Appl. Math. Comput. **285**, 79–93 (2016)
15. Derakhshandeh, S.Y., Pourbagher, R., Kargar, A.: A novel fuzzy logic Leven-berg-Marquardt method to solve the ill-conditioned power flow problem. Int. J. Electr. Power Energy Syst. **99**, 299–308 (2018)
16. Qiao, J., Wang, L., Yang, C., Gu, K.: Adaptive levenberg-marquardt algorithm based echo state network for chaotic time series prediction. IEEE Access **6**, 10720–10732 (2018)
17. Kumar, P., Sivalingam, S.M., Govindaraj, V.: Forecasting of HIV/AIDS in South Africa using 1990 to 2021 data: novel integer-and fractional-order fittings. Int. J. Dyn. Cont **12**(7), 2247–2263 (2023)
18. Hale, J.: Theory of Functional Differential Equations. Springer, New York (1997)
19. Kuang, Y.: Delay Differential Equations with Applications in Population Dynamics. Academic Press, Boston, Mathematics, Science, and Engineering (1993)
20. MacDonald, N.: Biological Delay Systems: Linear Stability Theory. Cambridge University, Cambridge (1989)